LONDON MATHEMATICAL SOCIETY LECTURE NOTE SERIES

Managing Editor: Professor N.J. Hitchin, Mathematical Institute,
University of Oxford, 24–29 St Giles, Oxford OX1 3LB, United Kingdom

The titles below are available from booksellers, or, in case of difficulty, from Cambridge University Press.

London Mathematical Society Lecture Note Series. 272

Character Theory
for the Odd Order Theorem

Thomas Peterfalvi
Université de Paris VII

Translated by Robert Sandling
University of Manchester

CAMBRIDGE
UNIVERSITY PRESS

CAMBRIDGE UNIVERSITY PRESS
Cambridge, New York, Melbourne, Madrid, Cape Town,
Singapore, São Paulo, Delhi, Mexico City

Cambridge University Press
The Edinburgh Building, Cambridge CB2 8RU, UK

Published in the United States of America by Cambridge University Press, New York

www.cambridge.org
Information on this title: www.cambridge.org/9780521646604

First published in French by Astérisque as *Théorie des charactéres dans le théoreme de Feit et Thompson* and *Le Théorem de Bender-Suzuki II.*

English edition first published 2000

A catalogue record for this publication is available from the British Library

Library of Congress Cataloguing in Publication Data
Peterfalvi, Thomas.
Character theory for the odd order theorem / Thomas Peterfalvi;
translated by R. Sandling.
    p.   cm. – (London Mathematical Society Lecture Note Series; 272)
Includes bibliographical references and index.
ISBN 0 521 64660 X (pbk.)
1. Feit Thompson theorem. 2. Finite groups. 3. Characters of groups. I. Title.
II. Series.
QA177.P48   1999
511'.2–dc21   99–25752   CIP

ISBN 978-0-521-64660-4 Paperback

London Mathematical Society Lecture Note Series. 272

Character Theory
for the Odd Order Theorem

Thomas Peterfalvi
Université de Paris VII

Translated by Robert Sandling
University of Manchester

CAMBRIDGE UNIVERSITY PRESS
Cambridge, New York, Melbourne, Madrid, Cape Town,
Singapore, São Paulo, Delhi, Mexico City

Cambridge University Press
The Edinburgh Building, Cambridge CB2 8RU, UK

Published in the United States of America by Cambridge University Press, New York

www.cambridge.org
Information on this title: www.cambridge.org/9780521646604

First published in French by Astérisque as *Théorie des charactéres dans le théoreme de Feit
et Thompson* and *Le Théorem de Bender-Suzuki II.*

English edition first published 2000

A catalogue record for this publication is available from the British Library

Library of Congress Cataloguing in Publication Data
Peterfalvi, Thomas.
Character theory for the odd order theorem / Thomas Peterfalvi;
translated by R. Sandling.
     p.   cm. – (London Mathematical Society Lecture Note Series; 272)
Includes bibliographical references and index.
ISBN 0 521 64660 X (pbk.)
1. Feit Thompson theorem. 2. Finite groups. 3. Characters of groups. I. Title.
II. Series.
QA177.P48   1999
511'.2–dc21   99–25752   CIP

ISBN 978-0-521-64660-4 Paperback

Contents

Part II A Theorem of Suzuki

Preface

This book comprises two parts, each devoted to the revision of the proof of a theorem about finite groups. The two theorems are among the results taken as foundation material in the revision of the classification of finite simple groups undertaken by D. Gorenstein, R. Lyons and R. Solomon.

The famous theorem of W. Feit and J. G. Thompson states that every finite group of odd order is solvable. It is the most important among the initial theorems in the classification of finite simple groups. The proof of this theorem divides into two parts. The first consists of the study of the maximal subgroups of a minimal counterexample to the theorem. This part, which is of considerable difficulty, has been revised by H. Bender and G. Glauberman; their work has appeared as a book in this series. The second part of the proof of the Feit-Thompson Theorem uses character theory to show that the existence of a simple group of odd order is impossible. In Part I of this book, we give a revision of this portion of the proof. Thus, with the book of Bender and Glauberman, a complete proof of the theorem is provided.

In Part II of this book, a revised proof of a theorem of M. Suzuki is given. This theorem characterizes certain groups which have a split BN-pair of rank 1. Let G be a 2-transitive group of permutations of a set X of odd order. Assume that the stabilizer in G of a point x of X has a normal subgroup acting regularly on $X - \{x\}$, and that subgroups of G fixing two points of X have odd order. The theorem shows that, if G is simple, then G is a group of Lie type of rank 1 in characteristic 2. This part is based on an earlier version which appeared, in French, in 1986.

I wish to record my warmest gratitude to Professor G. Glauberman who has both encouraged me to undertake, and assisted me in effecting, the preparation of each of these texts. I am also very grateful to R. Sandling who has carried out the translation into English with great care and who has suggested various improvements.

Part I
Character Theory
for the Odd Order Theorem

Introduction

The Feit-Thompson Theorem states that every finite group of odd order is solvable. This statement is clearly equivalent to the following: there is no non-abelian simple group of odd order. The theorem, first conjectured by Burnside, was proved in 1963 by W. Feit and J. G. Thompson in [FT]. Two papers, which prove the theorem in special cases, preceded the appearance of [FT]. In [Su], M. Suzuki proved the theorem for CA-groups of odd order: a group G is a *CA-group* if, for every element $x \neq 1$ of G, $C_G(x)$ is abelian. In [FHT], the theorem was shown for the CN-groups of odd order: a group G is a *CN-group* if, for every element $x \neq 1$ of G, $C_G(x)$ is nilpotent.

Each of these proofs is divided into two parts. In the first part, a minimal counterexample G to the theorem is considered and the structure of the maximal subgroups of G is studied. This part is very short in [Su], but is much more complicated in [FHT], and considerably more so in [FT]. In the second part, a contradiction is obtained by the use of character theory. The existence of isometries between virtual characters of maximal subgroups of G and virtual characters of G is one of the basic tools. In [FT], this second part leaves a residual case in which no contradiction arises. This case is eliminated in the final chapter of [FT], by explicit calculations with relations between elements of G.

The object of the present monograph is a revision of the second part of the proof of the Feit-Thompson Theorem, which corresponds to Chapters III and V of [FT].

From its appearance, [FT] has been the object of several efforts at revision. In [B], H. Bender gave a new proof of the Uniqueness Theorem, one of the principal results of Chapter IV of [FT]. From 1975, G. Glauberman worked on the revision of the first part of the proof of the theorem. In unpublished work [Si2], D. A. Sibley revised almost completely the part concerning characters. In [Pe], a revision of Chapter VI of [FT] was published by the present author. Finally, in 1994, H. Bender and G. Glauberman published a complete revision [BG] of the first part.

The present work may be viewed as a continuation of [BG] and constitutes with that book a complete proof of the Feit-Thompson Theorem. It is possible, however, to read this text without having read [BG] as the results of [BG] are reviewed in § 8.

We assume that the reader has a basic knowledge of ordinary character theory. There are many books which provide this theory. Here the book [Is] of I. M. Isaacs is used as reference. More precisely, the results of [Is] which are assumed known, are as follows:

Chapters 1 and 2;

in Chapter 3, (3.1) to (3.7), (3.11), (3.14);

in Chapter 4, (4.1), (4.2), (4.20), (4.21);

in Chapter 5, (5.1) to (5.5), (5.7) to (5.9);

in Chapter 6, (6.1) to (6.8), (6.10), (6.11), (6.28) (which uses the results (6.16) to (6.20) and (6.24) to (6.27)), (6.32) to (6.34);

in Chapter 7, (7.1) to (7.7).

We also assume known the following result from Problem 2.2 of [Is]:

Let G be a finite group, $|G| = n$, $\chi \in \operatorname{Irr} G$, σ be an automorphism of $\mathbf{Q}_n$ and χ^σ the mapping from G to $\mathbf{C}$ defined by $\chi^\sigma(g) = \chi(g)^\sigma$ for $g \in G$. Then $\chi^\sigma \in \operatorname{Irr} G$.

A certain familiarity with the elementary theory of finite groups is assumed. For the results used in this subject, reference is made to the initial sections of [BG] in so far as is possible. We also use Theorem 12.4 of [HB], Chapter XI, and Satz 8.18 of [H], Kapitel V.

The text is divided into sections. Sections 1 to 7 contain preliminary results. Sections 8 to 14 study a minimal counterexample to the Feit-Thompson Theorem. The hypotheses and results of § 29, for example, would be numbered (29.1), (29.2), ... Intermediate results used in the proof of (29.3) would be numbered (29.3.1), (29.3.2), ... If (29.4) is followed by a statement whose status is not specified, this statement is a lemma or a proposition.

I wish to thank Professor G. Glauberman, who suggested that I write this text and who helped me with advice and by supplying me with the relevant literature. The work of D. A. Sibley was of great utility to me in preparing this text.

Notation

Let G be a finite group.

We denote inclusion in the broad sense by $\subset$ (e.g., $G \subset G$).

$\mathrm{Irr}(G)$ or $\mathrm{Irr}\,G$ is the set of *irreducible characters* of G over the field $\mathbf{C}$.

$\mathrm{CF}(G)$ is the set of *class functions* from G to $\mathbf{C}$.

If $\alpha,\ \beta \in \mathrm{CF}(G)$, $(\alpha, \beta)_G$ or (α, β) is the usual *scalar product* of α and β, and $\|\alpha\|^2 = (\alpha, \alpha)$.

If $\phi \in \mathrm{CF}(G)$, $\mathrm{Supp}(\phi) = \{x \in G \mid \phi(x) \neq 0\}$.

Let $A \subset G$. Then $\mathrm{CF}(G, A) = \{\phi \in \mathrm{CF}(G) \mid \mathrm{Supp}(\phi) \subset A\}$.

If $\mathcal{X} \subset \mathrm{CF}(G)$ and R is a subring of $\mathbf{C}$, $R[\mathcal{X}]$ or $R\mathcal{X}$ is the set of R-linear combinations of elements of $\mathcal{X}$, and $R[\mathcal{X}, A] = R[\mathcal{X}] \cap \mathrm{CF}(G, A)$.

A *virtual character* of G is an element of $\mathbf{Z}[\mathrm{Irr}\,G]$.

If H is a subgroup of G, Res^G_H is *restriction*, $\mathrm{CF}(G) \to \mathrm{CF}(H)$, and Ind^G_H is *induction*, $\mathrm{CF}(H) \to \mathrm{CF}(G)$.

The symbol 1_G denotes the *principal character* of G. For $\chi \in \mathrm{CF}(G)$, $\overline{\chi}$ is defined by $\overline{\chi}(g) = \overline{\chi(g)}$ for $g \in G$.

If H is a normal subgroup of G, $\theta \in \mathrm{CF}(H)$ and $g \in G$, then θ^g is defined by $\theta^g(x^g) = \theta(x)$ for all $x \in H$. If $\theta \in \mathrm{Irr}(H)$, $I(\theta)$ or $I_G(\theta)$ is the *inertia group* of θ in G, which is the set of $g \in G$ such that $\theta^g = \theta$.

If $n \in \mathbf{N}$, $\mathbf{Q}_n$ is the subfield of $\mathbf{C}$ generated by the nth roots of unity.

A subset A of G is a *TI-subset* of G if, for every $g \in G$, $A^g = A$ or $A^g \cap A = \emptyset$.

If $A \subset G$, $A^{\#} = A - \{1\}$.

If $A \subset G$ and $L \subset G$, $A^L = \{a^x \mid a \in A, x \in L\}$.

We denote the *exponent* of G by $\exp(G)$. This is the smallest integer $n \geq 1$ such that $g^n = 1$ for every $g \in G$.

The symbol $\pi(G)$ denotes the set of prime divisors of the order $|G|$ of G.

Let σ be a set of prime numbers. We say that G is a *σ-group* if $\pi(G) \subset \sigma$. We denote by σ' the set of prime numbers which do not belong to σ. If $g \in G$, we denote by g_σ and $g_{\sigma'}$ the elements of $\langle g \rangle$ such that $g = g_\sigma g_{\sigma'} = g_{\sigma'} g_\sigma$ and $\pi(\langle g_\sigma \rangle) \subset \sigma$, $\pi(\langle g_{\sigma'} \rangle) \subset \sigma'$. We denote the largest normal σ-subgroup of G by $O_\sigma(G)$. If $\sigma = \{p\}$, we set $g_p = g_\sigma$, $g_{p'} = g_{\sigma'}$, $O_p(G) = O_\sigma(G)$ and $O_{p'}(G) = O_{\sigma'}(G)$.

$F(G)$ is the largest normal nilpotent subgroup of G.

$\Phi(G)$ is the *Frattini subgroup* of G.

The notation $G = H \rtimes K$ means that H and K are subgroups of G, H is normal in G, $G = HK$ and $H \cap K = 1$.

A group H *acts fixed-point-freely* on G, or *without fixed points* on G, if H acts on G and, for $g \in G$ and $h \in H$, $g^h = g$ implies that $h = 1$ or $g = 1$.

If p is a prime number and G is a p-group, $\Omega_1(G)$ is the subgroup of G generated by the elements of G of order p.

1. Preliminary Results from Character Theory

Let G be a finite group.

(1.1) *If $|G|$ is odd, $\chi \in \mathrm{Irr}(G)$ and $\chi \neq 1_G$, then $\overline{\chi} \neq \chi$.*

Proof. Let $\langle a \rangle$ be a group of order 2, and set $g^a = g^{-1}$ and $\chi^a = \overline{\chi}$ for $g \in G$ and $\chi \in \mathrm{Irr}(G)$. Then $\langle a \rangle$ acts on the set G and on $\mathrm{Irr}(G)$, and $\chi^a(g^a) = \chi(g)$ for all $g \in G$ and $\chi \in \mathrm{Irr}(G)$. By [Is], Theorem 6.32, the number of $\chi \in \mathrm{Irr}(G)$ such that $\overline{\chi} = \chi$ is then equal to the number of conjugacy classes C of G such that $C^{-1} = C$. Let C be such a class and let $g \in C$. There is an element $x \in G$ such that $g^{-1} = g^x$. It follows that x^2 centralizes g. But x is of odd order, and so $x \in \langle x^2 \rangle \subset C_G(g)$, whence $g^{-1} = g$, and, since g is of odd order, $g = 1$. The only class C such that $C^{-1} = C$ is $\{1\}$, and so 1_G is the only character $\chi \in \mathrm{Irr}(G)$ such that $\overline{\chi} = \chi$. $\qquad\square$

(1.2) *Let H be a normal subgroup of G, let $\chi \in \mathrm{Irr}(G)$ be such that $H \not\subset \mathrm{Ker}\,\chi$ and let $g \in G$ be such that $C_H(g) = 1$. Then $\chi(g) = 0$.*

Proof. Let $\overline{g}$ be the image of g in G/H. Since $C_H(g) = 1$, we have $|C_G(g)| = |C_G(g)H/H| \leq |C_{G/H}(\overline{g})|$. If Φ is the set of irreducible characters of G which have H in their kernels, then, by the second orthogonality relation,

$$\sum_{\chi \in \mathrm{Irr}(G)} |\chi(g)|^2 = |C_G(g)| \leq |C_{G/H}(\overline{g})| = \sum_{\phi \in \Phi} |\phi(g)|^2.$$

Thus $\chi(g) = 0$ if $\chi \in \mathrm{Irr}(G) - \Phi$. $\qquad\square$

(1.3) *Let H be a subgroup of G and let A be a union of conjugacy classes of H. Let $(\psi_j)_{j \in J}$ be a basis for $\mathrm{CF}(H, A)$ and let $\mathrm{Irr}(H) = \{\chi_i \mid i \in I\}$.*

(a) Let $\mu \in \mathrm{CF}(G)$ and $d_i \in \mathbf{C}$ for $i \in I$. Then $\mu_{|A} = (\sum_{i \in I} d_i \chi_i)_{|A}$ if and only if, for all $j \in J$, $\sum_{i \in I} (\psi_j, \chi_i)_H \overline{d_i} = (\mathrm{Ind}_H^G \psi_j, \mu)_G$.

(b) Suppose that there is an orthonormal family $(\mu_i)_{i \in I}$ of elements of $\mathrm{CF}(G)$ such that, for all $j \in J$, $\mathrm{Ind}_H^G \psi_j = \sum_{i \in I} (\psi_j, \chi_i) \mu_i$. Then $\mu_{i|A} = \chi_{i|A}$ for all $i \in I$, and, if μ is orthogonal to μ_i for all $i \in I$, $\mu_{|A} = 0$.

Proof. (a) Considering the orthogonal basis of $\mathrm{CF}(H)$ consisting of the characteristic functions of conjugacy classes, we see that the orthogonal complement of $\mathrm{CF}(H, A)$ in $\mathrm{CF}(H)$ is $\mathrm{CF}(H, H - A)$. It follows that

$$\mathrm{Res}_H^G \mu - \sum_{i \in I} d_i \chi_i \in \mathrm{CF}(H, H - A)$$

is equivalent to

$$\text{for all } j \in J, \ \left(\psi_j, \mathrm{Res}_H^G \mu - \sum_{i \in I} d_i \chi_i\right) = 0;$$

that is, to

$$\text{for all } j \in J, \ \sum_{i \in I} (\psi_j, \chi_i)\overline{d_i} = (\psi_j, \mathrm{Res}_H^G \mu).$$

The result follows by Frobenius reciprocity.

(b) We apply (a) with $\mu = \mu_i$, $d_i = 1$ and $d_k = 0$ for $k \neq i$, and then with μ orthogonal to μ_i and $d_i = 0$ for all i. $\qquad\square$

(1.4) *Let H be a finite group, let $\mathcal{X} = \{\chi_1, \ldots, \chi_n\} \subset \mathrm{Irr}(H)$ be such that $|\mathcal{X}| = n \geq 2$ and $\chi_i(1) = \chi_1(1)$ for all i, $1 \leq i \leq n$, and let τ be an isometry from $\mathbf{Z}[\mathcal{X}, H^\#]$ to $\mathbf{Z}[\mathrm{Irr}\, G, G^\#]$. Then there are pairwise distinct characters $\mu_i \in \mathrm{Irr}(G)$ and an integer $\varepsilon = \pm 1$ such that $(\chi_i - \chi_1)^\tau = \varepsilon(\mu_i - \mu_1)$ for all i, $1 \leq i \leq n$.*

Proof. It suffices to show that there is an orthonormal family (e_i) of $\mathbf{Z}[\mathrm{Irr}\, G]$ such that $(\chi_i - \chi_1)^\tau = e_i - e_1$ for all i. In fact, there are then characters $\mu_i \in \mathrm{Irr}\, G$ and integers $\varepsilon_i = \pm 1$ such that $e_i = \varepsilon_i \mu_i$, while $(\chi_i - \chi_1)^\tau(1) = \varepsilon_i \mu_i(1) - \varepsilon_1 \mu_1(1) = 0$ implies that $\varepsilon_i = \varepsilon_1$. If $n = 2$, $\|(\chi_2 - \chi_1)^\tau\|^2 = 2$, and so $(\chi_2 - \chi_1)^\tau$ can be written as $e_2 - e_1$. If $n = 3$, $(\chi_2 - \chi_1)^\tau$ and $(\chi_3 - \chi_1)^\tau$ are sums of two orthogonal components of norm 1. Since $((\chi_2 - \chi_1)^\tau, (\chi_3 - \chi_1)^\tau) = 1$, they have a component in common, with the same sign, which we can write $-e_1$, and so we obtain the required decompositions. Suppose that $n > 3$ and that $(\chi_i - \chi_1)^\tau = e_i - e_1$ for $2 \leq i < k$, where $3 < k \leq n$. On taking into account the fact that $((\chi_k - \chi_1)^\tau, (\chi_i - \chi_1)^\tau) = 1$ for $i < k$, we see that one of two cases holds:

$$(\chi_k - \chi_1)^\tau = e_k - e_1 \quad \text{or} \quad (\chi_k - \chi_1)^\tau = e_2 + e_3.$$

But, in the second case, $(e_2 + e_3)(1) = (e_2 - e_3)(1) = 0$, and so $e_2(1) = 0$, which is a contradiction. $\qquad\square$

(1.5) *Let H be a normal subgroup of G, $\theta \in \mathrm{Irr}(H)$, $r = |I_G(\theta) : H|$ and $\chi = \mathrm{Ind}_H^G \theta$.*

(a) *$\mathrm{Res}_H^G \chi = r \sum \theta^g$, where the sum is taken over the $|G : I_G(\theta)|$ distinct conjugates of θ in G.*

(b) *$\|\chi\|^2 = r$. In particular, χ is irreducible if $r = 1$.*

(c) *Let $\phi \in \mathrm{Irr}(H)$. If ϕ is conjugate to θ in G, then $\mathrm{Ind}_H^G \phi = \chi$, and, if ϕ is not conjugate to θ, then $(\mathrm{Ind}_H^G \phi, \chi) = 0$.*

(d) *$\dfrac{\chi(1)\mathrm{Res}_H^G \chi}{\|\chi\|^2} = |G : H| \sum \theta^g(1)\theta^g$, where the sum is indexed as in (a).*

(e) *If $|G|$ is odd and $\theta \neq 1_H$, then $\overline{\chi}$ is orthogonal to χ.*

Proof. (a) For $h \in H$, $\chi(h) = \dfrac{1}{|H|} \sum_{x \in G} \theta(xhx^{-1}) = \dfrac{1}{|H|} \sum_{x \in G} \theta^x(h)$. But $\theta^x = \theta^y$ if and only if $y \in I(\theta)x$, and the result follows.

(b) By (a) and Frobenius reciprocity, $\|\chi\|^2 = (\mathrm{Res}_H^G \chi, \theta) = r$.

(c) If ϕ is conjugate to θ, then (a) shows that $\mathrm{Ind}_H^G \phi$ and χ coincide on H. As they vanish on $G - H$, they coincide on G. If ϕ is not conjugate to θ, then $(\mathrm{Ind}_H^G \phi, \chi) = (\phi, \mathrm{Res}_H^G \chi) = 0$ by (a).

(d) Using (a) and (b), we see that

$$\frac{\chi(1)\operatorname{Res}_H^G \chi}{\|\chi\|^2} = \frac{|G:H|\theta(1)r \sum \theta^g}{r} = |G:H| \sum \theta^g(1)\theta^g.$$

(e) First of all, $\overline{\chi} = \operatorname{Ind}_H^G \overline{\theta}$. If $\overline{\chi}$ is not orthogonal to χ, then, by (c), there is an element $g \in G$ such that $\overline{\theta} = \theta^g$. Then $\theta^{g^2} = \theta$. But $g \in \langle g^2 \rangle$ because $|G|$ is odd, and so $\overline{\theta} = \theta^g = \theta$. By (1.1), it follows that $\theta = 1_H$. $\qquad\square$

(1.6) *Let H be a normal subgroup of G, let $\theta \in \operatorname{Irr}(H)$ and let A be a normal subgroup of G contained in H.*

(a) *$A \subset \operatorname{Ker}\theta$ if and only if $A \subset \operatorname{Ker}\operatorname{Ind}_H^G \theta$.*

(b) *Suppose that $A \subset \operatorname{Ker}\theta$. Let θ_1 be the character of H/A for which $\theta_1(xA) = \theta(x)$ for all $x \in H$. Let χ be the character of G/A for which $\chi(xA) = (\operatorname{Ind}_H^G \theta)(x)$ for all $x \in G$. Then $\chi = \operatorname{Ind}_{H/A}^{G/A} \theta_1$.*

Proof. (a) This follows from (1.5.a) and from [Is], Lemma 2.21.

(b) We see that χ and $\operatorname{Ind}_{H/A}^{G/A} \theta_1$ vanish on $G/A - H/A$. For $h \in H$,

$$(\operatorname{Ind}_{H/A}^{G/A} \theta_1)(hA) = \frac{|A|}{|H|} \sum_{xA \in G/A} \theta_1(xhAx^{-1})$$

$$= \frac{1}{|H|} \sum_{x \in G} \theta(xhx^{-1}) = \chi(hA).$$

$$\square$$

(1.7) *Let H be a normal subgroup of G, $\theta \in \operatorname{Irr}(H)$ and $T = I_G(\theta)$. Set $\operatorname{Ind}_H^T \theta = \sum_{i=1}^n e_i\psi_i$, where $e_i \in \mathbf{N} - \{0\}$ and the characters ψ_i are distinct elements of $\operatorname{Irr}(T)$.*

(a) *Let $\chi_i = \operatorname{Ind}_T^G \psi_i$ $(1 \le i \le n)$. Then the characters χ_i are distinct elements of $\operatorname{Irr}(G)$; furthermore, $\operatorname{Ind}_H^G \theta = \sum_{i=1}^n e_i\chi_i$.*

(b) *Suppose that T/H is abelian. Then $\operatorname{Ind}_H^G \theta = e \sum_{i=1}^n \chi_i$, where $e = e_1$, $n = |T:H|/e^2$ and $\chi_i(1) = |G:T|e\theta(1)$ for all i $(1 \le i \le n)$.*

(c) *Suppose that T/H is abelian and that $|H|$ is prime to the index $|T:H|$. Then $\operatorname{Ind}_H^G \theta = \sum_{i=1}^n \chi_i$, $n = |T:H|$ and, for all i, $\chi_i(1) = |G:T|\theta(1)$.*

Proof. (a) This follows from [Is], Theorem 6.11.

(b) Let $L = \{\lambda \in \operatorname{Irr} T \mid H \subset \operatorname{Ker}\lambda\}$. Since T/H is abelian, $\sum_{\lambda \in L} \lambda$ is the regular character of T/H, identified with a character of T. With $\psi = \psi_1$, $\sum_{\lambda \in L}(\lambda\psi)(x) = |T:H|\psi(x)$ for $x \in H$ and $\sum_{\lambda \in L}(\lambda\psi)(x) = 0$ for $x \in T - H$. It follows that $\operatorname{Ind}_H^T(\operatorname{Res}_H^T \psi) = \sum_{\lambda \in L} \lambda\psi$. Moreover, $\lambda\psi \in \operatorname{Irr} T$ for $\lambda \in L$, because $\lambda(1) = 1$. Since θ is a component of $\operatorname{Res}_H^T \psi$, this proves that the characters ψ_i are of the form $\lambda\psi$, $\lambda \in L$. Furthermore,

$$(\operatorname{Ind}_H^T \theta, \lambda\psi) = (\theta, \operatorname{Res}_H^T(\lambda\psi)) = (\theta, \operatorname{Res}_H^T \psi),$$

and so $\mathrm{Ind}_H^T \theta = e \sum_{i=1}^n \psi_i$, where $e = e_1$. By Clifford's Theorem ([Is], Theorem 6.5), $\mathrm{Res}_H^T \psi = e\theta$, and so, for all i, $\psi_i(1) = \psi(1) = e\theta(1)$, whence

$$|T : H|\theta(1) = (\mathrm{Ind}_H^T \theta)(1) = ne^2\theta(1).$$

Thus $n = |T : H|/e^2$. By (a), it follows that

$$\mathrm{Ind}_H^G \theta = e \sum_{i=1}^n \chi_i \quad \text{and} \quad \chi_i(1) = |G : T|\psi(1) = |G : T|e\theta(1).$$

(c) By [Is], Corollary 6.28, there is an index i such that $e_i = 1$; (b) then gives the desired result. $\square$

(1.8) *Let $\psi \in \mathrm{Irr}\, G$ and let B, C and D be subgroups of G. Assume that B is normal in C, that $B \subset \mathrm{Ker}\,\psi$, that $B \subset D \subset C$, and that $D/B \subset Z(C/B)$. Then $\psi(1) \leq |G|/\sqrt{|C||D|}$.*

Proof. Let χ be an irreducible component of $\mathrm{Res}_C^G \psi$. Then $B \subset \mathrm{Ker}\,\chi$ and, by [Is], Corollary 2.30, $\chi(1)^2 \leq |C : D|$. Since ψ is an irreducible component of $\mathrm{Ind}_C^G \chi$,

$$\psi(1)^2 \leq |G : C|^2\chi(1)^2 \leq \frac{|G|^2}{|C||D|}.$$

$\square$

(1.9) *Suppose that $|G| = n = ab$, where a and b are relatively prime.*

(a) *Let u be an automorphism of the field $\mathbf{Q}_a$. There is an automorphism v of $\mathbf{Q}_n$ such that $v_{|\mathbf{Q}_a} = u$ and $v_{|\mathbf{Q}_b} = \mathrm{Id}$.*

(b) *Let χ be a character of G and let $k \in \mathbf{Z}$ be such that $(k, a) = 1$. There is an automorphism v of $\mathbf{Q}_n$ such that*

$$\chi^v(g) = \begin{cases} \chi(g^k) & \text{if the order of } g \text{ divides } a, \\ \chi(g) & \text{if the order of } g \text{ is prime to } a. \end{cases}$$

Proof. (a) We may assume that $a, b \geq 2$. Let m be an integer ≥ 2. There is a group isomorphism $f_m : (\mathbf{Z}/m\mathbf{Z})^* \to \mathrm{Aut}(\mathbf{Q}_m)$ such that, if $\phi_m : \mathbf{Z} \to \mathbf{Z}/m\mathbf{Z}$ is the canonical mapping and ε is an mth root of unity, then $f_m(\phi_m(k))(\varepsilon) = \varepsilon^k$ for k prime to m ([L], Chapter VIII, Theorem 3.1). Let

$$i : (\mathbf{Z}/n\mathbf{Z})^* \to (\mathbf{Z}/a\mathbf{Z})^* \times (\mathbf{Z}/b\mathbf{Z})^*$$

be the isomorphism given by the Chinese Remainder Theorem. Let $h = (f_a \times f_b) \circ i \circ f_n^{-1}$. Let $v = f_n(\phi_n(k))$, an automorphism of $\mathbf{Q}_n$. Then $h(v) = (f_a(\phi_a(k)), f_b(\phi_b(k)))$ and, if ε is an nth root of unity,

$$f_a(\phi_a(k))(\varepsilon^b) = f_n(\phi_n(k))(\varepsilon^b) = \varepsilon^{bk}.$$

Thus $h(v) = (v_{|\mathbf{Q}_a}, v_{|\mathbf{Q}_b})$, and the result follows since h is bijective.

(b) Let $u = f_a(\phi_a(k))$ and let v be the automorphism of $\mathbf{Q}_n$ given by (a). For $g \in G$, the restriction of χ to $\langle g \rangle$ is of the form $\sum_{i=1}^{\chi(1)} \psi_i$, where the ψ_i are homomorphisms from $\langle g \rangle$ to $\mathbf{C}^*$. If the order of g divides a, then $\psi_i(g) \in \mathbf{Q}_a$ and so

$$\chi^v(g) = \sum \psi_i(g)^k = \sum \psi_i(g^k) = \chi(g^k)$$

while, if the order of g is prime to a, then $\psi_i(g) \in \mathbf{Q}_b$ and so $\chi^v(g) = \chi(g)$. $\square$

(1.10) *Let p be a prime number, ε a primitive pth root of unity in $\mathbf{C}$, η a primitive $|G|$th root of unity in $\mathbf{C}$ and $A = \mathbf{Z}[\eta]$.*

(a) *Let x, $y \in G$ be such that x has order p and $xy = yx$. Let χ be a virtual character of G. Then $\chi(xy) \equiv \chi(y)$ (mod $1 - \varepsilon$) in A.*

(b) *If $n \in \mathbf{Z}$ and $n \equiv 0$ (mod $1 - \varepsilon$) in A, then $n \equiv 0$ (mod p) in $\mathbf{Z}$.*

Proof. (a) Let α be an irreducible component of $\mathrm{Res}^G_{\langle x,y \rangle} \chi$. It suffices to show that $\alpha(xy) \equiv \alpha(y)$ (mod $1 - \varepsilon$). Since $xy = yx$, α is of degree 1, and so $\alpha(xy) - \alpha(y) = (\alpha(x) - 1)\alpha(y)$. There is an integer $k \geq 1$ such that $\alpha(x) = \varepsilon^k$. Thus, $\alpha(x) - 1 = (\varepsilon - 1)(\varepsilon^{k-1} + \cdots + 1)$ is divisible by $1 - \varepsilon$ in A. Therefore $\alpha(xy) - \alpha(y)$ is divisible by $1 - \varepsilon$ in A.

(b) If $1 \leq k < p$, there is an integer $r \geq 1$ for which $\varepsilon = \varepsilon^{kr}$, and so $\varepsilon - 1 = (\varepsilon^k - 1)(\varepsilon^{k(r-1)} + \cdots + 1)$ and $1 - \varepsilon^k$ divides n. If $F(X) = X^{p-1} + X^{p-2} + \cdots + 1$, then

$$F(X) = \prod_{1 \leq k < p} (X - \varepsilon^k) \quad \text{and} \quad F(1) = \prod_{1 \leq k < p} (1 - \varepsilon^k) = p.$$

Thus p divides n^{p-1} in A. By [Is], Lemma 3.2 and Corollary 3.5, $A \cap \mathbf{Q} = \mathbf{Z}$, and so p divides n^{p-1} in $\mathbf{Z}$, and, since p is prime, p divides n. $\square$

2. The Dade Isometry

Let G be a finite group, let A be a TI-subset of G and let $L = N_G(A)$. By [Is], Lemma 7.7, Ind_L^G is a linear isometry from $\operatorname{CF}(L, A)$ to $\operatorname{CF}(G)$, which sends each virtual character of $\operatorname{CF}(L, A)$ to a virtual character of G. We give here a generalization of this isometry when A satisfies a condition less restrictive than that of being a TI-subset.

(2.1) *Let G be a finite group. Let $g \in G$ and let H be a subgroup of G such that g normalizes H and such that $\langle g \rangle$ and H have coprime orders. Then Hg is the disjoint union of $|H : C_H(g)|$ subsets which are conjugate to $C_H(g)g$ in $H\langle g \rangle$.*

Proof. Let $K = \bigcup_{x \in H}(C_H(g)g)^x$. Then $K \subset Hg$ since $H \lhd H\langle g \rangle$. Let π be the set of prime divisors of the order of g. Let $x, y \in H$ and $u, v \in C_H(g)$ be such that $(ug)^x = (vg)^y$. Then $g^x = ((ug)^x)_\pi = ((vg)^y)_\pi = g^y$, and so $y \in C_H(g)x$. It follows that K is the disjoint union of $(C_H(g)g)^x$ for x running through a system of right coset representatives of $C_H(g)$ in H. We conclude that

$$|K| = |H : C_H(g)||C_H(g)g| = |H| = |Hg|,$$

which proves that $K = Hg$. $\qquad\square$

(2.2) Hypothesis. *Assume that G is a finite group, that A is a subset of $G^{\#}$ and that L is a subgroup of G such that $A \subset L \subset N_G(A)$. Furthermore:*

   (a) *If two elements of A are conjugate in G, then they are conjugate in L.*

   (b) *For $a \in A$, there is a subgroup $H(a)$ of G such that*

$$C_G(a) = H(a) \rtimes C_L(a).$$

   (c) *For $a, b \in A$, $|H(a)|$ is prime to $|C_L(b)|$.*

(2.3) *Let A be a non-empty subset of $G^{\#}$. Then A is a TI-subset of G with normalizer L if and only if Hypothesis (2.2) holds with $H(a) = 1$ for all $a \in A$.*

Proof. Suppose that A is a TI-subset of G, with normalizer L. Let $a \in A$ and $g \in G$ be such that $a^g \in A$. Then $a^g \in A \cap A^g$ and so $g \in N_G(A) = L$. In particular, if $g \in C_G(a)$, then $g \in L$ and so $C_G(a) = C_L(a)$. Thus, Hypothesis (2.2) holds with $H(a) = 1$ for all $a \in A$. Suppose conversely that Hypothesis (2.2) holds with $H(a) = 1$ for all $a \in A$. Let $g \in G$ be such that $A \cap A^g \neq \emptyset$. Let a be such that $a^g \in A \cap A^g$. By (2.2.a), there is an element $x \in L$ such that $a^{gx} = a$, and so $gx \in C_G(a) = C_L(a)$, whence $g \in L \subset N_G(A)$. Thus A is a TI-subset of G. Let $g \in N_G(A)$. Since $A \neq \emptyset$, $A \cap A^g \neq \emptyset$ and we have seen that then $g \in L$. Thus $N_G(A) = L$. $\qquad\square$

   In the remainder of § 2, we will assume Hypothesis (2.2).

(2.4) (a) *Let $a \in A$ and $x \in L$. Then $H(a^x) = H(a)^x$.*

(b) *Let a, $b \in A$ be such that $(aH(a))^G \cap (bH(b))^G \neq \emptyset$. Then a and b are conjugate in L.*

(c) *For $a \in A$, $N_G(aH(a)) = C_G(a)$.*

Proof. Let π be the union of the sets of prime divisors of $|C_L(a)|$ for $a \in A$. By (2.2.b,c), $H(a) = O_{\pi'}(C_G(a))$ for all $a \in A$. Thus, if $a \in A$ and $x \in L$, then $H(a^x) = H(a)^x$. Let a, $b \in A$, $u \in H(a)$ and $v \in H(b)$ be such that au is conjugate to bv in G. Then $a = (au)_\pi$ is conjugate to $b = (bv)_\pi$ in G, and so a is conjugate to b in L by (2.2.a). If $a \in A$ and $g \in N_G(aH(a))$, then $a^g = (a^g)_\pi = a$, and so $g \in C_G(a)$. $\qquad\square$

(2.5) Definition. Let $\alpha \in \mathrm{CF}(L, A)$. We define a class function α^τ on G by setting $\alpha^\tau(g) = \alpha(a)$ if g is conjugate in G to an element of $aH(a)$ for some $a \in A$, and $\alpha^\tau(g) = 0$ if $g \notin \bigcup_{a \in A}(aH(a))^G$.

By (2.4.b), $\alpha^\tau(g)$ is well defined. In this section, we prove the following theorem.

(2.6) Theorem. (a) *If α, $\beta \in \mathrm{CF}(L, A)$, then $(\alpha^\tau, \beta^\tau)_G = (\alpha, \beta)_L$.*

(b) *If $\alpha \in \mathbf{Z}[\mathrm{Irr}\, L, A]$, then $\alpha^\tau \in \mathbf{Z}[\mathrm{Irr}\, G]$.*

We say that τ is the *Dade isometry* relative to (A, L, G), or relative to A.

(2.7) *Let $\alpha \in \mathrm{CF}(L, A)$ and $\chi \in \mathrm{CF}(G)$. Let $\psi \in \mathrm{CF}(L)$ be such that, for all $a \in A$,*

$$\psi(a) = \frac{1}{|H(a)|} \sum_{x \in H(a)} \chi(ax).$$

Then $(\alpha^\tau, \chi)_G = (\alpha, \psi)_L$. In particular, if χ is constant on $aH(a)$ for all $a \in A$, then $(\alpha^\tau, \chi)_G = (\alpha, \mathrm{Res}_L^G \chi)_L$.

Proof. Let T be a system of representatives for the conjugacy classes of L contained in A. On taking (2.4) into account, we see that

$$
\begin{aligned}
(\alpha^\tau, \chi)_G &= \frac{1}{|G|} \sum_{g \in G} \alpha^\tau(g)\overline{\chi}(g) = \frac{1}{|G|} \sum_{a \in T} \sum_{g \in (aH(a))^G} \alpha^\tau(g)\overline{\chi}(g) \\
&= \frac{1}{|G|} \sum_{a \in T} \alpha(a) \frac{|G|}{|C_G(a)|} \sum_{x \in H(a)} \overline{\chi}(ax) = \sum_{a \in T} \frac{\alpha(a)}{|C_L(a)|} \overline{\psi}(a) \\
&= \frac{1}{|L|} \sum_{a \in T} |a^L| \alpha(a)\overline{\psi}(a) \\
&= (\alpha, \psi)_L.
\end{aligned}
$$

$$\square$$

(2.8) *Let B be a non-empty subset of A. Set $H(B) = \bigcap_{a \in B} H(a)$ and set $M(B) = H(B)N_L(B)$. Then $M(B) = H(B) \rtimes N_L(B)$.*

Proof. By (2.4.a), $N_L(B)$ normalizes $H(B)$. For $a \in B$,

$$H(B) \cap N_L(B) \subset H(a) \cap L = H(a) \cap C_L(a) = 1.$$

Thus $M(B) = H(B) \rtimes N_L(B)$. $\hspace{2cm}$ $\square$

(2.9) Notation. Let $\alpha \in \mathrm{CF}(L, A)$ and let B be a non-empty subset of A. We define $\alpha_B \in \mathrm{CF}(M(B))$ by $\alpha_B(hx) = \alpha(x)$ for $h \in H(B)$ and $x \in N_L(B)$. Thus, $\alpha_B = \alpha \circ f_B$, where f_B is the natural homomorphism $M(B) \to L$ with kernel $H(B)$. It follows that, if α is a virtual character of L, then α_B is a virtual character of $M(B)$.

(2.10) *Let $\mathcal{B}$ be a system of L-conjugacy class representatives of the non-empty subsets of A. If $\alpha \in \mathrm{CF}(L, A)$, then*

$$\alpha^\tau = - \sum_{B \in \mathcal{B}} (-1)^{|B|} \mathrm{Ind}^G_{M(B)} \alpha_B.$$

(2.10.1) *Let $x \in L$ and let B be a non-empty subset of A. Then*

$$\mathrm{Ind}^G_{M(B^x)} \alpha_{B^x} = \mathrm{Ind}^G_{M(B)} \alpha_B.$$

Proof. By (2.4.a),

$$M(B^x) = H(B^x) \rtimes N_L(B^x) = H(B)^x \rtimes N_L(B)^x = M(B)^x.$$

Furthermore, $\alpha_{B^x}(h^x g^x) = \alpha(g^x) = \alpha(g) = \alpha_B(hg)$ for $h \in H(B)$ and for $g \in N_L(B)$. The result follows by the definition of induction. $\hspace{1cm}$ $\square$

(2.10.2) *Let B be a non-empty subset of A and let $a \in A$. Then*

$$C_{H(B)}(a) = H(B \cup \{a\}).$$

Proof. As $H(B \cup \{a\}) \subset H(a) \subset C_G(a)$, we see that $H(B \cup \{a\}) \subset C_{H(B)}(a)$. Conversely, $C_{H(B)}(a)$ is a subgroup of $C_G(a)$ of order prime to $|C_L(a)|$. Thus $C_{H(B)}(a) \subset H(a)$ and $C_{H(B)}(a) \subset H(a) \cap H(B) = H(B \cup \{a\})$. $\hspace{1cm}$ $\square$

If $g \in G$ and $X \subset G$, set $\mathcal{A}(g, X) = \{x \in G \mid g^x \in X\}$.

(2.10.3) *Let B be a non-empty subset of A. If $g \notin \bigcup_{a \in A}(aH(a))^G$, then* $(\mathrm{Ind}^G_{M(B)} \alpha_B)(g) = 0$. *If $g \in (aH(a))^G$ for some $a \in A$, then*

$$(\mathrm{Ind}^G_{M(B)} \alpha_B)(g) = \frac{\alpha(a)}{|M(B)|} \sum_{b \in N_L(B) \cap (a^L)} |\mathcal{A}(g, H(B)b)|.$$

Proof. We know that

$$(\mathrm{Ind}^G_{M(B)} \alpha_B)(g) = \frac{1}{|M(B)|} \sum_{x \in \mathcal{A}(g, M(B))} \alpha_B(x^{-1} g x).$$

Let $x \in \mathcal{A}(g, M(B))$ be such that $\alpha_B(x^{-1}gx) \neq 0$. Then $x^{-1}gx = hb$, where $h \in H(B)$, $b \in N_L(B)$ and $\alpha(b) \neq 0$. Thus $b \in A$. It then follows from (2.1) and from (2.2.c) that hb is conjugate to an element of $C_{H(B)}(b)b$, and so $g \in (bH(b))^G$. If $g \notin \bigcup_{a \in A}(aH(a))^G$, then $(\mathrm{Ind}^G_{M(B)}\alpha_B)(g) = 0$. Suppose that $g \in (aH(a))^G$ for some $a \in A$. Then, by (2.4.b), $b \in N_L(B) \cap (a^L)$. Moreover, $\alpha_B(x^{-1}gx) = \alpha(b) = \alpha(a)$, which gives the required conclusion. $\qquad\square$

Proof of (2.10). Let $\gamma = -\sum_{B \in \mathcal{B}}(-1)^{|B|}\mathrm{Ind}^G_{M(B)}\alpha_B$ and let $g \in G$. If $g \notin \bigcup_{a \in A}(aH(a))^G$, then $\alpha^\tau(g) = \gamma(g) = 0$ by (2.10.3). Suppose then that $g \in (aH(a))^G$ for some $a \in A$. Let $\mathcal{P}$ be the set of non-empty subsets of A. By (2.10.1) and (2.10.3),

$$\gamma(g) \;=\; -\sum_{B \in \mathcal{P}} \frac{(-1)^{|B|}}{|L : N_L(B)|}(\mathrm{Ind}^G_{M(B)}\alpha_B)(g)$$

$$\;=\; -\alpha(a)\sum_{B \in \mathcal{P}} \frac{(-1)^{|B|}}{|L : N_L(B)||M(B)|}\sum_{b \in N_L(B) \cap (a^L)} |\mathcal{A}(g, H(B)b)|.$$

If $\mathcal{P}(b)$ is the set of $B \in \mathcal{P}$ such that $b \in N_L(B)$, then

$$\gamma(g) = -\alpha(a)\sum_{b \in a^L}\sum_{B \in \mathcal{P}(b)} \frac{(-1)^{|B|}}{|L||H(B)|}|\mathcal{A}(g, H(B)b)|.$$

If $x \in L$ and $b = a^x$, then $B \in \mathcal{P}(a)$ is equivalent to $B^x \in \mathcal{P}(b)$, and also $y \in \mathcal{A}(g, H(B)a)$ is equivalent to $yx \in \mathcal{A}(g, H(B^x)b)$. Thus $|\mathcal{A}(g, H(B^x)b)| = |\mathcal{A}(g, H(B)a)|$ and

$$\gamma(g) = -\alpha(a)|a^L|\sum_{B \in \mathcal{P}(a)} \frac{(-1)^{|B|}}{|L||H(B)|}|\mathcal{A}(g, H(B)a)|$$

$$= \frac{-\alpha(a)}{|C_L(a)|}\sum_{B \in \mathcal{P}(a)} \frac{(-1)^{|B|}}{|H(B)|}|\mathcal{A}(g, H(B)a)|.$$

By (2.1), if $B \in \mathcal{P}(a)$, then $H(B)a$ is the disjoint union of $|H(B) : C_{H(B)}(a)|$ subsets conjugate to $C_{H(B)}(a)a$. Thus,

$$|\mathcal{A}(g, H(B)a)| = |\mathcal{A}(g, C_{H(B)}(a)a)||H(B) : C_{H(B)}(a)|$$

and

$$\gamma(g) = \frac{-\alpha(a)}{|C_L(a)|}\sum_{B \in \mathcal{P}(a)} \frac{(-1)^{|B|}}{|C_{H(B)}(a)|}|\mathcal{A}(g, C_{H(B)}(a)a)|.$$

Let B be a non-empty subset of $A - \{a\}$. Then $B \in \mathcal{P}(a)$ is equivalent to $B \cup \{a\} \in \mathcal{P}(a)$ and, if $B \in \mathcal{P}(a)$, $C_{H(B)}(a) = C_{H(B \cup \{a\})}(a) = H(B \cup \{a\})$ by (2.10.2). In the sum above, terms cancel two at a time, aside from that for which $B = \{a\}$. By hypothesis, $\mathcal{A}(g, H(a)a)$ is non-empty, and, if x is in this set, then $\mathcal{A}(g, H(a)a) = xC_G(a)$. Therefore,

$$\gamma(g) = \frac{-\alpha(a)}{|C_L(a)|}\frac{(-1)}{|H(a)|}|\mathcal{A}(g, H(a)a)| = \alpha(a)\frac{|C_G(a)|}{|C_L(a)||H(a)|} = \alpha(a).$$

Thus $\gamma = \alpha^\tau$. $\qquad\qquad\qquad\qquad\qquad\qquad\qquad\qquad\qquad\qquad\qquad\qquad\square$

Proof of Theorem (2.6). (a) By definition, β^τ is constant on $aH(a)$, for $a \in A$, and so, by (2.7),

$$(\alpha^\tau, \beta^\tau)_G = (\alpha, \mathrm{Res}_L^G \beta^\tau)_L = \frac{1}{|L|}\sum_{a \in A}\alpha(a)\overline{\beta^\tau(a)} = \frac{1}{|L|}\sum_{a \in A}\alpha(a)\overline{\beta(a)} = (\alpha, \beta)_L.$$

(b) This is a consequence of (2.10). $\qquad\qquad\qquad\qquad\qquad\qquad\qquad\square$

(2.11) *Let A_1 be a subset of A normalized by L. Then Hypothesis (2.2) holds with A_1 in place of A. If τ_1 is the Dade isometry relative to A_1, then τ_1 is the restriction of τ to* $\mathrm{CF}(L, A_1)$.

Proof. It is clear that Hypothesis (2.2) holds with A_1 in place of A. Let $\alpha \in \mathrm{CF}(L, A_1)$. Then $\alpha^\tau(g) = \alpha(a)$ if g is conjugate to an element of $aH(a)$ for some $a \in A$, and $\alpha^\tau(g) = 0$ otherwise. Thus, $\alpha^\tau(g) = \alpha(a)$ if g is conjugate to an element of $aH(a)$ for some $a \in A_1$, and $\alpha^\tau(g) = 0$ otherwise. Consequently, $\alpha^\tau = \alpha^{\tau_1}$. $\qquad\qquad\qquad\qquad\qquad\qquad\qquad\qquad\qquad\qquad\qquad\qquad\square$

3. TI-Subsets with Cyclic Normalizers

In this section, we will assume

(3.1) Hypothesis. *G is a finite group and $W = W_1 \times W_2$ is a subgroup of G. Assume that W is cyclic of odd order, that $w_1 = |W_1| \neq 1$ and $w_2 = |W_2| \neq 1$, and that $V = W - (W_1 \cup W_2)$ is a TI-subset of G with normalizer W.*

Since W is cyclic, w_1 and w_2 are relatively prime. We know that Ind_W^G is a linear isometry from $\mathrm{CF}(W, V)$ to $\mathrm{CF}(G)$ which sends virtual characters to virtual characters ([Is], Chapter 7).

In this section, we prove the following theorem.

(3.2) Theorem. *There is a linear isometry $\sigma : \mathrm{CF}(W) \to \mathrm{CF}(G)$ which sends virtual characters of W to virtual characters of G, such that:*

(a) *If $\alpha \in \mathrm{CF}(W, V)$, then $\alpha^\sigma = \mathrm{Ind}_W^G \alpha$.*

(b) *$1_W^\sigma = 1_G$.*

(c) *If $\alpha \in \mathrm{CF}(W)$ and $x \in V$, then $\alpha^\sigma(x) = \alpha(x)$.*

(d) *Every irreducible character of G which is not in the image of σ vanishes on V.*

(3.3) Notation. Let $\omega_{00} = 1_W$, let ω_{i0} $(0 \leq i < w_1)$ be the irreducible characters of W which have W_2 in their kernels, let ω_{0j} $(0 \leq j < w_2)$ be the irreducible characters of W which have W_1 in their kernels, and let $\omega_{ij} = \omega_{i0}\omega_{0j}$.

Each ω_{ij} has degree 1 and $\mathrm{Irr}(W) = \{\omega_{ij} \mid 0 \leq i < w_1, 0 \leq j < w_2\}$ by [Is], Corollary 2.23 and Theorem 4.21.

(3.4) *Let $\alpha_{ij} = 1_W - \omega_{i0} - \omega_{0j} + \omega_{ij}$. Then the family (α_{ij}), $1 \leq i < w_1$ and $1 \leq j < w_2$, is a basis of the $\mathbf{C}$-space $\mathrm{CF}(W, V)$.*

Proof. Since $\alpha_{ij} = (1_W - \omega_{i0})(1_W - \omega_{0j})$, we see that $\alpha_{ij} \in \mathrm{CF}(W, V)$. Let $a_{ij} \in \mathbf{C}$ be such that $\sum a_{ij}\alpha_{ij} = 0$ $(1 \leq i < w_1, 1 \leq j < w_2)$. Then $a_{ij} = (\sum a_{kl}\alpha_{kl}, \omega_{ij}) = 0$ for $1 \leq i < w_1, 1 \leq j < w_2$, and so the functions α_{ij} are linearly independent. But

$$\dim \mathrm{CF}(W, V) = |V| = |W| - |W_1| - |W_2| + 1 = (w_1 - 1)(w_2 - 1),$$

and so the result follows. $\qquad\square$

(3.5) *There is an orthonormal family (χ_{ij}) $(0 \leq i < w_1, 0 \leq j < w_2)$ of elements of $\mathbf{Z}\mathrm{Irr}(G)$ such that*

$$\chi_{00} = 1_G \quad \text{and} \quad \mathrm{Ind}_W^G \alpha_{ij} = 1_G - \chi_{i0} - \chi_{0j} + \chi_{ij}$$

for $i, j \geq 1$.

Assume that (3.5) has been shown. Set $\omega_{ij}^{\sigma} = \chi_{ij}$ and extend σ to CF(W) by linearity. Then (a) and (b) of Theorem (3.2) are established, and assertions (c) and (d) of Theorem (3.2) follow from (1.3).

Proof of (3.5).

(3.5.1) *Let* $\beta_{ij} = \mathrm{Ind}_W^G \alpha_{ij} - 1_G$ $(1 \le i < w_1, 1 \le j < w_2)$. *Then* $(\beta_{ij}, 1_G) = 0$ *and* $\|\beta_{ij}\|^2 = 3$ *for all* i, j *while* $(\beta_{ij}, \beta_{ij'}) = (\beta_{ij}, \beta_{i'j}) = 1$ *and* $(\beta_{ij}, \beta_{i'j'}) = 0$ *for* $i \ne i', j \ne j'$.

Proof. That $(\mathrm{Ind}_W^G \alpha_{ij}, 1_G) = (\alpha_{ij}, 1_W) = 1$ follows from Frobenius reciprocity, and so $(\beta_{ij}, 1_G) = 0$. The other relations follow from the fact that Ind_W^G is an isometry on CF(W, V). $\qquad\square$

Let $1 \le i < w_1, 1 \le j < w_2$. By (3.5.1) and the fact that $\beta_{ij} \in \mathbf{Z}\mathrm{Irr}(G)$, we see that $\beta_{ij} = \sum_{\chi \in A_{ij}} \chi$, where A_{ij} is a set of three pairwise orthogonal elements of $\pm(\mathrm{Irr}(G) - \{1_G\})$.

(3.5.2) *We have* $|A_{11} \cap A_{12}| = 1$ *and* $A_{11} \cap (-A_{12}) = \emptyset$.

Proof. Let $A_{11} = \{\chi_1, \chi_2, \chi_3\}$ and $a_i = (\beta_{12}, \chi_i)$ for $i = 1, 2, 3$. Then $(\beta_{12}, \beta_{11}) = a_1 + a_2 + a_3 = 1$ and $a_i \in \{0, 1, -1\}$. The numbers a_i are thus either $1, 0, 0$, or $1, 1, -1$. In the second case, we may assume that $\beta_{12} = \chi_1 + \chi_2 - \chi_3$ whence $2\chi_3 = \beta_{11} - \beta_{12} = \mathrm{Ind}_W^G(\alpha_{11} - \alpha_{12})$ vanishes on $1 \in G$, which is a contradiction. $\qquad\square$

Lemma (3.5.2) clearly holds with A_{ij} and $A_{i'j'}$ in place of A_{11} and A_{12} if $i = i'$ and $j \ne j'$ or if $i \ne i'$ and $j = j'$. We refer to this lemma for A_{ij} and $A_{i'j'}$ as L$(ij, i'j')$. We also refer to the statement $(\beta_{ij}, \beta_{i'j'}) = 0$ for $i \ne i'$ and $j \ne j'$ as O$(ij, i'j')$.

By Hypothesis (3.1), $\sup(w_1, w_2) \ge 5$. By the symmetry between w_1 and w_2, we will assume

(3.5.3) $w_1 \ge 5$.

In the proof which follows, the functions χ_i and χ_{ij} are pairwise orthogonal elements of $\pm(\mathrm{Irr}(G) - \{1_G\})$.

(3.5.4) $|\bigcap_{1 \le i < w_1} A_{i1}| = 1$.

Proof. Suppose that (3.5.4) is false. By (3.5.2), we can then write, for some choice of indices $i = 1, 2, 3$,

$$\begin{aligned}
\beta_{11} &= \chi_1 + \chi_2 + \chi_3, \\
\beta_{21} &= \chi_1 + \chi_4 + \chi_5, \\
\beta_{31} &= \chi_2 + \chi_4 + \chi_6.
\end{aligned}$$

We consider two cases:

Case I. There are indices i and i' such that $1 \leq i < i' \leq 3$ and
$$A_{i1} \cap A_{i'1} \cap A_{41} \neq \emptyset.$$

Case II. For $1 \leq i < i' < i'' \leq 4$, $A_{i1} \cap A_{i'1} \cap A_{i''1} = \emptyset$.

Suppose that Case I holds. Up to choice of notation, we have, by (3.5.2),

$$\beta_{41} = \chi_1 + \chi_6 + \chi_7.$$

(3.5.4.1)

If $\chi_1 \in A_{12}$, then $\beta_{12} = \chi_1 - \chi_5 - \chi_7$.

If $\chi_1 \in A_{22}$, then $\beta_{22} = \chi_1 - \chi_3 - \chi_7$.

If $\chi_1 \in A_{42}$, then $\beta_{42} = \chi_1 - \chi_3 - \chi_5$.

Proof. Suppose that $\chi_1 \in A_{12}$. By O(12, 21), $-\chi_4$ or $-\chi_5 \in A_{12}$. Suppose that $-\chi_4 \in A_{12}$. Then it follows from O(12, 31) and L(12, 11) that $\chi_6 \in A_{12}$, which contradicts O(12, 41). Thus $-\chi_5 \in A_{12}$. Similarly, we see that $-\chi_7 \in A_{12}$ by interchanging the roles of β_{21} and β_{41}.

The other two assertions follow from the symmetry between β_{11}, β_{21} and β_{41}. $\qquad\square$

(3.5.4.2) *We may assume that $\beta_{32} = \chi_2 - \chi_3 + \chi_8$.*

Proof. By the symmetry between the functions β_{11}, β_{21} and β_{41}, we may assume that $A_{32} \cap A_{31} = \{\chi_2\}$. By O(32, 11), $-\chi_1$ or $-\chi_3 \in A_{32}$. Suppose that $-\chi_1 \in A_{32}$. Then, by O(32, 21) and O(32, 41), the third element of A_{32} is in $A_{21} \cap A_{41}$, which is a contradiction. Thus $-\chi_3 \in A_{32}$ and the third element of A_{32} cannot be one of the functions $\pm\chi_i$, $i \leq 7$. $\qquad\square$

(3.5.4.3) *We may assume that $\beta_{12} = \chi_2 - \chi_4 + \chi_5$.*

Proof. By (3.5.4.1), (3.5.4.2) and L(12, 32), χ_1 does not belong to A_{12}. By L(12, 11), χ_2 or $\chi_3 \in A_{12}$. But, by L(12, 32) and (3.5.4.2), $\chi_3 \notin A_{12}$. Thus $\chi_2 \in A_{12}$. By O(12, 31), $-\chi_4$ or $-\chi_6 \in A_{12}$. By the symmetry between the functions β_{21} and β_{41}, we may assume that $-\chi_4 \in A_{12}$. Then $\chi_5 \in A_{12}$ by O(12, 21). $\qquad\square$

(3.5.4.4) $\beta_{22} = \chi_5 + \chi_8 + \chi_9$.

Proof. By (3.5.4.1), (3.5.4.3) and L(22, 12), χ_1 does not belong to A_{22}. By L(22, 21), χ_4 or $\chi_5 \in A_{22}$. Then L(22, 12) shows that $\chi_5 \in A_{22}$. By L(22, 12), $\chi_2 \notin A_{22}$ and so L(22, 32) implies that $-\chi_3$ or $\chi_8 \in A_{22}$. But, by O(22, 11), $-\chi_3 \notin A_{22}$ and so $\chi_8 \in A_{22}$. Thus the third element of A_{22} cannot be one of the functions $\pm\chi_i$, $i \leq 8$. $\qquad\square$

(3.5.4.5) *Case I is impossible.*

Proof. By (3.5.4.1) and L(42, 12), χ_1 does not belong to A_{42}. Suppose that χ_2 or $-\chi_3 \in A_{42}$. By O(42, 11), χ_2 and $-\chi_3 \in A_{42}$, which, with (3.5.4.2) taken into account, contradicts L(42, 32). Thus $\chi_2 \notin A_{42}$ and $-\chi_3 \notin A_{42}$. By L(42, 32), $\chi_8 \in A_{42}$. By L(42, 22), we have $\chi_5 \notin A_{42}$ and so, by L(42, 12), $-\chi_4 \in A_{42}$. But this contradicts O(42, 21). $\square$

(3.5.4.6) *Case II is impossible.*

Proof. Suppose that Case II holds. By (3.5.2),

$$\beta_{41} = \chi_3 + \chi_5 + \chi_6.$$

We may assume that $\chi_1 \in A_{12}$ by L(12, 11) and the symmetry between the functions χ_1, χ_2 and χ_3. We may assume that $-\chi_4 \in A_{12}$ by O(12, 21) and the symmetry between the functions χ_4 and χ_5. Then O(12, 31) implies that $\chi_6 \in A_{12}$ and O(12, 41) is contradicted. $\square$

(3.5.5) *The following decompositions hold for* $1 \leq i < w_1$:

$$\begin{aligned}
\beta_{i1} &= -\chi_{i0} - \chi_{01} + \chi_{i1}, \\
\beta_{i2} &= -\chi_{i0} - \chi_{02} + \chi_{i2}.
\end{aligned}$$

Proof. Let $-\chi_{01}$ be the element of $\bigcap_i A_{i1}$. Suppose that $\chi_{01} \in \pm A_{12}$. Then $O(12, i1)$ for $i > 1$ implies that A_{12} contains an element of $\pm A_{i1}$ distinct from $\pm \chi_{01}$. Thus $|A_{12}| \geq 1 + (w_1 - 2) \geq 4$, which is a contradiction. Thus χ_{01} is orthogonal to A_{i2}. Let $-\chi_{02}$ be the element of $\bigcap_i A_{i2}$. Again, χ_{02} is orthogonal to A_{i1}, and we can denote the element of $A_{i1} \cap A_{i2}$ by $-\chi_{i0}$. If we set $\beta_{i1} = -\chi_{i0} - \chi_{01} + \phi_i$ and $\beta_{i2} = -\chi_{i0} - \chi_{02} + \psi_i$, (3.5.2) shows that (ϕ_i) and (ψ_i) are orthonormal families, orthogonal to $-\chi_{i0}$, to χ_{01} and to χ_{02}. By $L(i1, i2)$ and $O(i1, i'2)$ for $i' \neq i$, the functions ϕ_i are orthogonal to the functions $\psi_{i'}$. $\square$

If $w_2 = 3$, the proof of (3.5) is complete. Suppose that $w_2 \geq 5$. Then (3.5.4) holds with W_1 and W_2 interchanged, and so $-\chi_{i0} \in A_{ij}$ for all j. By applying (3.5.5) to β_{i1} and β_{ij} for some fixed $j \geq 2$, then to β_{ij} and $\beta_{ij'}$ for two arbitrary indices j and j', we see that we can write $\beta_{ij} = -\chi_{i0} - \chi_{0j} + \chi_{ij}$ for $1 \leq i < w_1$, $1 \leq j < w_2$. This completes the proof of (3.5). $\square$

(3.6) Hypothesis. *Let* σ *be as in Theorem* (3.2), *let* ω_{ij} *be as in* (3.3) *and let* $\psi \in \mathrm{CF}(G)$. *Set*

$$\psi = \sum_{\substack{0 \leq i < w_1 \\ 0 \leq j < w_2}} a_{ij}\omega_{ij}^{\sigma} + \beta,$$

where $a_{ij} \in \mathbf{C}$ *and* $\beta \in \mathrm{CF}(G)$ *is orthogonal to the image of* σ. *Assume that* ψ *vanishes on* V. *Set* $NC(\psi) = |\{(i, j) \mid 0 \leq i < w_1, 0 \leq j < w_2, a_{ij} \neq 0\}|$.

(3.7) *Assume Hypothesis* (3.6). *Let* i, i' *and* j, j' *be indices which satisfy* $0 \leq i, i' < w_1$, $0 \leq j, j' < w_2$. *Then* $a_{ij} + a_{i'j'} = a_{ij'} + a_{i'j}$. *In particular,* $a_{ij} = a_{i0} + a_{0j} - a_{00}$.

Proof. Let $\alpha = \omega_{ij} + \omega_{i'j'} - \omega_{ij'} - \omega_{i'j}$. We see that $\text{Supp}(\alpha) \subset V$, and so $\text{Supp}(\text{Ind}_W^G \alpha) \subset V^G$. As $\text{Supp}(\psi) \cap V^G = \emptyset$, it follows that $(\psi, \text{Ind}_W^G \alpha) = a_{ij} + a_{i'j'} - a_{ij'} - a_{i'j} = 0$. $\qquad\square$

(3.8) *Assume Hypothesis* (3.6), *that* $w_1 < w_2$ *and that* $NC(\psi) < 2w_1$. *Then one of the following three cases holds.*

(a) $\psi = \beta$.

(b) $NC(\psi) = w_1$ *and there is an element* $a \in \mathbf{C}$ *and an index* j *such that*

$$\psi = a \sum_{0 \le i < w_1} \omega_{ij}^\sigma + \beta.$$

(c) $NC(\psi) = w_2$ *and there is an element* $a \in \mathbf{C}$ *and an index* i *such that*

$$\psi = a \sum_{0 \le j < w_2} \omega_{ij}^\sigma + \beta.$$

Proof. Suppose that $0 < NC(\psi) < 2w_1$. We need only use (3.7), in which there is a symmetry among the indices i (and among the indices j as well). We may then assume that $a_{00} \ne 0$. In what follows, i, j and k are indices such that $0 \le i < w_1$, $0 \le j, k < w_2$.

(3.8.1) *If* $i \ge 1$ *and* $j \ge 1$, *then* $a_{i0} \ne 0$ *or* $a_{0j} \ne 0$.

Proof. Suppose, for example, that $a_{10} = a_{01} = 0$. By (3.7), $a_{11} = -a_{00} \ne 0$. For $i \ge 2$, $a_{i0} = a_{i1} + a_{00}$ and so $a_{i0} \ne 0$ or $a_{i1} \ne 0$. For $j \ge 2$, $a_{0j} = a_{1j} + a_{00}$ and so $a_{0j} \ne 0$ or $a_{1j} \ne 0$. Thus $NC(\psi) \ge 2 + (w_1 - 2) + (w_2 - 2) = w_1 + w_2 - 2 \ge 2w_1$, which contradicts the hypothesis. $\qquad\square$

(3.8.2) *Suppose that there is an index* $j \ge 1$ *such that* $a_{0j} = 0$. *Then case* (b) *holds.*

Proof. Suppose, for example, that $a_{01} = 0$. By (3.8.1), $a_{i0} \ne 0$ for all $i \ge 0$. Suppose that there is an index $j \ge 2$ such that $a_{0j} \ne 0$. Interchanging the roles of 0 and j, we then see that $a_{ij} \ne 0$ for all $i \ge 0$. But then $NC(\psi) \ge 2w_1$, contrary to the hypothesis. Thus $a_{0j} = 0$ for all $j \ge 1$. Suppose that there are indices $i \ge 1$ and $j \ge 1$ such that $a_{ij} \ne 0$, for example, that $a_{11} \ne 0$. Then, for $j \ge 2$,

$$a_{1j} = a_{1j} + a_{01} = a_{11} + a_{0j} = a_{11} \ne 0.$$

Thus $NC(\psi) \ge w_1 + (w_2 - 1) > 2w_1$, which contradicts the hypothesis. Thus $a_{ij} = 0$ for all $j \ge 1$. By (3.7), $a_{i0} = a_{00}$ for all i and $\psi = a_{00} \sum_i \omega_{i0}^\sigma + \beta$. $\qquad\square$

(3.8.3) *Suppose that* $a_{0j} \ne 0$ *for all* j. *Then case* (c) *holds.*

Proof. Suppose that there are indices $i \ge 1$ and j such that $a_{ij} \ne 0$. If there is an index k such that $a_{ik} = 0$, then (3.8.2), with the roles of (i, j) and $(0, 0)$ interchanged, shows that case (b) holds, which contradicts the hypothesis of

(3.8.3). Thus $a_{ik} \neq 0$ for all k and $NC(\psi) \geq 2w_2 > 2w_1$, which contradicts the hypothesis. Thus $a_{ij} = 0$ for all $i \geq 1$. By (3.7), $a_{0j} = a_{00}$ for all j and $\psi = a_{00} \sum_j \omega_{0j}^\sigma + \beta$. $\qquad\qquad\qquad\qquad\qquad\qquad\qquad\qquad\qquad\qquad\qquad\square$

(3.9) *Let σ be as in Theorem (3.2) and let $\omega \in \mathrm{Irr}(W)$.*

   (a) *If $\chi \in \pm\mathrm{Irr}(G)$ is such that $\chi_{|V} = \omega_{|V}$, then $\chi = \omega^\sigma$. In particular, if u is an automorphism of the field $\mathbf{Q}_{|G|}$, then $\omega^{\sigma u} = \omega^{u\sigma}$.*

   (b) *Let a be the multiplicative order of ω and k an integer prime to a. There is an automorphism u of $\mathbf{Q}_{|G|}$ such that $(\omega^k)^\sigma = \omega^{\sigma u}$, and, if $g \in G$ has order prime to a, $(\omega^k)^\sigma(g) = \omega^\sigma(g)$.*

   (c) *Let a be the multiplicative order of ω and let $g \in G$ be of order prime to a. Then $\omega^\sigma(g) \in \mathbf{Z}$.*

Proof. (a) By (3.2.c) and the hypothesis, $\omega^\sigma - \chi$ vanishes on V; moreover, $NC(\omega^\sigma - \chi) \leq 2$. By (3.8), $\omega^\sigma - \chi$ is orthogonal to ω^σ, and so $\chi = \omega^\sigma$. The second assertion follows because $\omega^{\sigma u}{}_{|V} = \omega^u{}_{|V}$ by (3.2.c), and $\omega^{\sigma u} \in \pm\mathrm{Irr}(G)$.

   (b) By (1.9), there is an automorphism $u \in \mathrm{Aut}\,\mathbf{Q}_{|G|}$ such that

$$u(\varepsilon) = \begin{cases} \varepsilon^k & \text{if } \varepsilon \text{ is an } a\text{th root of unity,} \\ \varepsilon & \text{if } \varepsilon \text{ is a root of unity of order prime to } a. \end{cases}$$

Then $(\omega^k)^\sigma = \omega^{u\sigma} = \omega^{\sigma u}$, and, if g has order prime to a, $\omega^\sigma(g)$ is a sum of roots of unity of orders prime to a, and so $\omega^{\sigma u}(g) = \omega^\sigma(g)$.

   (c) Let $|G| = bc$, where b and c are relatively prime and the prime divisors of b are the prime divisors of the order of g. Let $u \in \mathrm{Aut}\,\mathbf{Q}_b$. By (1.9), there is an automorphism $v \in \mathrm{Aut}\,\mathbf{Q}_{|G|}$ such that $v_{|\mathbf{Q}_b} = u$ and $v_{|\mathbf{Q}_c} = \mathrm{Id}_{\mathbf{Q}_c}$. Since ω takes values in $\mathbf{Q}_c$ and $\omega^\sigma(g) \in \mathbf{Q}_b$,

$$\omega^\sigma(g)^u = \omega^{\sigma v}(g) = \omega^{v\sigma}(g) = \omega^\sigma(g).$$

By Galois theory, it follows that $\omega^\sigma(g) \in \mathbf{Q}$, and, as $\omega^\sigma(g)$ is an algebraic integer, $\omega^\sigma(g) \in \mathbf{Z}$. $\qquad\qquad\qquad\qquad\qquad\qquad\qquad\qquad\qquad\square$

Proof. Let $\alpha = \omega_{ij} + \omega_{i'j'} - \omega_{ij'} - \omega_{i'j}$. We see that $\operatorname{Supp}(\alpha) \subset V$, and so $\operatorname{Supp}(\operatorname{Ind}_W^G \alpha) \subset V^G$. As $\operatorname{Supp}(\psi) \cap V^G = \emptyset$, it follows that $(\psi, \operatorname{Ind}_W^G \alpha) = a_{ij} + a_{i'j'} - a_{ij'} - a_{i'j} = 0$. $\qquad\square$

(3.8) *Assume Hypothesis (3.6), that $w_1 < w_2$ and that $NC(\psi) < 2w_1$. Then one of the following three cases holds.*

  (a) $\psi = \beta$.

  (b) *$NC(\psi) = w_1$ and there is an element $a \in \mathbf{C}$ and an index j such that*

$$\psi = a \sum_{0 \le i < w_1} \omega_{ij}^\sigma + \beta.$$

  (c) *$NC(\psi) = w_2$ and there is an element $a \in \mathbf{C}$ and an index i such that*

$$\psi = a \sum_{0 \le j < w_2} \omega_{ij}^\sigma + \beta.$$

Proof. Suppose that $0 < NC(\psi) < 2w_1$. We need only use (3.7), in which there is a symmetry among the indices i (and among the indices j as well). We may then assume that $a_{00} \neq 0$. In what follows, i, j and k are indices such that $0 \le i < w_1$, $0 \le j, k < w_2$.

(3.8.1) *If $i \ge 1$ and $j \ge 1$, then $a_{i0} \neq 0$ or $a_{0j} \neq 0$.*

Proof. Suppose, for example, that $a_{10} = a_{01} = 0$. By (3.7), $a_{11} = -a_{00} \neq 0$. For $i \ge 2$, $a_{i0} = a_{i1} + a_{00}$ and so $a_{i0} \neq 0$ or $a_{i1} \neq 0$. For $j \ge 2$, $a_{0j} = a_{1j} + a_{00}$ and so $a_{0j} \neq 0$ or $a_{1j} \neq 0$. Thus $NC(\psi) \ge 2 + (w_1 - 2) + (w_2 - 2) = w_1 + w_2 - 2 \ge 2w_1$, which contradicts the hypothesis. $\qquad\square$

(3.8.2) *Suppose that there is an index $j \ge 1$ such that $a_{0j} = 0$. Then case (b) holds.*

Proof. Suppose, for example, that $a_{01} = 0$. By (3.8.1), $a_{i0} \neq 0$ for all $i \ge 0$. Suppose that there is an index $j \ge 2$ such that $a_{0j} \neq 0$. Interchanging the roles of 0 and j, we then see that $a_{ij} \neq 0$ for all $i \ge 0$. But then $NC(\psi) \ge 2w_1$, contrary to the hypothesis. Thus $a_{0j} = 0$ for all $j \ge 1$. Suppose that there are indices $i \ge 1$ and $j \ge 1$ such that $a_{ij} \neq 0$, for example, that $a_{11} \neq 0$. Then, for $j \ge 2$,

$$a_{1j} = a_{1j} + a_{01} = a_{11} + a_{0j} = a_{11} \neq 0.$$

Thus $NC(\psi) \ge w_1 + (w_2 - 1) > 2w_1$, which contradicts the hypothesis. Thus $a_{ij} = 0$ for all $j \ge 1$. By (3.7), $a_{i0} = a_{00}$ for all i and $\psi = a_{00} \sum_i \omega_{i0}^\sigma + \beta$. $\qquad\square$

(3.8.3) *Suppose that $a_{0j} \neq 0$ for all j. Then case (c) holds.*

Proof. Suppose that there are indices $i \ge 1$ and j such that $a_{ij} \neq 0$. If there is an index k such that $a_{ik} = 0$, then (3.8.2), with the roles of (i, j) and $(0, 0)$ interchanged, shows that case (b) holds, which contradicts the hypothesis of

(3.8.3). Thus $a_{ik} \neq 0$ for all k and $NC(\psi) \geq 2w_2 > 2w_1$, which contradicts the hypothesis. Thus $a_{ij} = 0$ for all $i \geq 1$. By (3.7), $a_{0j} = a_{00}$ for all j and $\psi = a_{00} \sum_j \omega_{0j}^\sigma + \beta$. $\square$

(3.9) *Let σ be as in Theorem (3.2) and let $\omega \in \mathrm{Irr}(W)$.*

(a) *If $\chi \in \pm\mathrm{Irr}(G)$ is such that $\chi_{|V} = \omega_{|V}$, then $\chi = \omega^\sigma$. In particular, if u is an automorphism of the field $\mathbf{Q}_{|G|}$, then $\omega^{\sigma u} = \omega^{u\sigma}$.*

(b) *Let a be the multiplicative order of ω and k an integer prime to a. There is an automorphism u of $\mathbf{Q}_{|G|}$ such that $(\omega^k)^\sigma = \omega^{\sigma u}$, and, if $g \in G$ has order prime to a, $(\omega^k)^\sigma(g) = \omega^\sigma(g)$.*

(c) *Let a be the multiplicative order of ω and let $g \in G$ be of order prime to a. Then $\omega^\sigma(g) \in \mathbf{Z}$.*

Proof. (a) By (3.2.c) and the hypothesis, $\omega^\sigma - \chi$ vanishes on V; moreover, $NC(\omega^\sigma - \chi) \leq 2$. By (3.8), $\omega^\sigma - \chi$ is orthogonal to ω^σ, and so $\chi = \omega^\sigma$. The second assertion follows because $\omega^{\sigma u}{}_{|V} = \omega^u{}_{|V}$ by (3.2.c), and $\omega^{\sigma u} \in \pm\mathrm{Irr}(G)$.

(b) By (1.9), there is an automorphism $u \in \mathrm{Aut}\,\mathbf{Q}_{|G|}$ such that

$$u(\varepsilon) = \begin{cases} \varepsilon^k & \text{if } \varepsilon \text{ is an } a\text{th root of unity,} \\ \varepsilon & \text{if } \varepsilon \text{ is a root of unity of order prime to } a. \end{cases}$$

Then $(\omega^k)^\sigma = \omega^{u\sigma} = \omega^{\sigma u}$, and, if g has order prime to a, $\omega^\sigma(g)$ is a sum of roots of unity of orders prime to a, and so $\omega^{\sigma u}(g) = \omega^\sigma(g)$.

(c) Let $|G| = bc$, where b and c are relatively prime and the prime divisors of b are the prime divisors of the order of g. Let $u \in \mathrm{Aut}\,\mathbf{Q}_b$. By (1.9), there is an automorphism $v \in \mathrm{Aut}\,\mathbf{Q}_{|G|}$ such that $v_{|\mathbf{Q}_b} = u$ and $v_{|\mathbf{Q}_c} = \mathrm{Id}_{\mathbf{Q}_c}$. Since ω takes values in $\mathbf{Q}_c$ and $\omega^\sigma(g) \in \mathbf{Q}_b$,

$$\omega^\sigma(g)^u = \omega^{\sigma v}(g) = \omega^{v\sigma}(g) = \omega^\sigma(g).$$

By Galois theory, it follows that $\omega^\sigma(g) \in \mathbf{Q}$, and, as $\omega^\sigma(g)$ is an algebraic integer, $\omega^\sigma(g) \in \mathbf{Z}$. $\square$

4. The Dade Isometry
for a Certain Type of Subgroup

(4.1) *Let X be a finite group, let α, β, γ, $\delta \in \pm\mathrm{Irr}(X)$, and let u and v be non-zero real numbers. Assume that $(\alpha,\beta) = (\gamma,\delta) = (\alpha - \beta, u\gamma - v\delta) = 0$ and that $(\alpha - \beta)(1) = (u\gamma - v\delta)(1) = 0$. Then α, β, γ and δ are pairwise orthogonal.*

Proof. Suppose, for example, that $(\alpha, \gamma) \neq 0$. There is thus an integer $\varepsilon = \pm 1$ such that $\gamma = \varepsilon\alpha$, and $0 = (\alpha - \beta, u\gamma - v\delta) = u\varepsilon + v(\beta,\delta)$. Thus $(\beta,\delta) \neq 0$ and $v\delta = v(\beta,\delta)\beta = -u\varepsilon\beta$. Then $0 = (u\gamma - v\delta)(1) = u\varepsilon(\alpha + \beta)(1) = 2u\varepsilon\alpha(1)$, which is a contradiction. $\qquad\qquad\qquad\qquad\qquad\qquad\qquad\qquad\qquad\quad\square$

(4.2) Hypothesis. (a) *Let $L = K \rtimes W_1$ be a finite group, where $W_1 \neq 1$ is a cyclic Hall subgroup of L.*

(b) *There is a cyclic subgroup $W_2 \neq 1$ of K such that, for all $x \in W_1^{\#}$, $C_K(x) = W_2$.*

(c) *$W = W_1 \times W_2$ is of odd order.*

(4.3) Theorem. *Assume Hypothesis* (4.2).

(a) *$W - W_2$ is a TI-subset of L, with normalizer W, and Hypothesis* (3.1) *holds with L in place of G.*

(b) *We use the notation of Hypotheses* (3.1) *and* (3.3). *There are pairwise distinct characters $\mu_{ij} \in \mathrm{Irr}(L)$, $0 \leq i < w_1$, $0 \leq j < w_2$, and integers $\delta_j = \pm 1$ such that $\mathrm{Ind}_W^L(\omega_{ij} - \omega_{0j}) = \delta_j(\mu_{ij} - \mu_{0j})$. The isometry defined in Theorem* (3.2) *with L in place of G sends ω_{ij} to $\delta_j\mu_{ij}$.*

(c) *For $0 \leq i < w_1$, $0 \leq j < w_2$ and $x \in W - W_2$, $\mu_{ij}(x) = \delta_j\omega_{ij}(x)$. Every irreducible character of L which is not one of the characters μ_{ij} vanishes on $W - W_2$.*

(d) *For $0 \leq i < w_1$ and $0 \leq j < w_2$, $\mu_{ij}(1) \equiv \delta_j \pmod{w_1}$.*

Proof. Let $xy \in W - W_2$, where $x \in W_1^{\#}$ and $y \in W_2$. Let $g \in L$ be such that $(xy)^g \in W - W_2$. Then $x^g \in W_1$ and, since L/K is commutative, $x^g \in Kx$. It follows that $x^g = x$, and so $g \in C_L(x) = W$. This proves (a). If $i > 0$ and $j \geq 0$, then $\omega_{ij} - \omega_{0j} \in \mathrm{CF}(W, W - W_2)$. The number of these virtual characters is $(w_1 - 1)w_2 = |W - W_2|$, and they are linearly independent. Thus the functions $\omega_{ij} - \omega_{0j}$, for $i > 0$, $j \geq 0$, comprise a basis of $\mathrm{CF}(W, W - W_2)$. We know that Ind_W^L is an isometry on $\mathrm{CF}(W, W - W_2)$. By (1.4), there are characters $\mu_{ij} \in \mathrm{Irr}(L)$ and integers $\delta_j = \pm 1$ such that $\mathrm{Ind}_W^L(\omega_{ij} - \omega_{0j}) = \delta_j(\mu_{ij} - \mu_{0j})$, and, for j fixed, the characters μ_{ij} are pairwise distinct. For $j' \neq j$, $(\mu_{ij} - \mu_{0j}, \mu_{i'j'} - \mu_{0j'}) = 0$ for all i', and so, by (4.1), the characters μ_{ij} are pairwise distinct. It then follows from (1.3) that, for $x \in W - W_2$, $\delta_j\mu_{ij}(x) = \omega_{ij}(x)$ and that, if $\mu \in \mathrm{Irr}(L)$ is orthogonal to the characters μ_{ij}, then μ vanishes on $W - W_2$. By (3.9.a), the isometry of Theorem (3.2) then sends ω_{ij} to $\delta_j \mu_{ij}$. By (c),

$$\mathrm{Res}^L_{W_1} \mu_{ij} - \mathrm{Res}^W_{W_1} \delta_j \omega_{ij} \in \mathbf{Z}[\mathrm{Irr}\, W_1, \{1\}].$$

There is thus an integer a such that

$$\mathrm{Res}^L_{W_1} \mu_{ij} = \delta_j \mathrm{Res}^W_{W_1} \omega_{ij} + a\rho_{W_1},$$

where ρ_{W_1} is the regular character of W_1, and so $\mu_{ij}(1) = \delta_j + aw_1$. $\qquad\square$

(4.4) *In the notation of Theorem (4.3), the characters μ_{i0} $(0 \le i < w_1)$ are the irreducible characters of L whose kernels contain K. Furthermore, $\delta_0 = 1$ and $\mu_{00} = 1_L$.*

Proof. Let $\chi \in \mathrm{Irr}(L)$ be such that $K \subset \mathrm{Ker}\,\chi$. There is thus an index i, $0 \le i < w_1$, such that $\mathrm{Res}^L_W \chi = \omega_{i0}$. By (3.9), it follows that $\chi = \delta_0 \mu_{i0}$. Since χ and μ_{i0} are characters, $\delta_0 = 1$. If $\chi = 1_L$, then $i = 0$ and so $\mu_{00} = 1_L$. Conversely, $K \subset \mathrm{Ker}\,\mu_{i0}$ for all i because $|\mathrm{Irr}(L/K)| = w_1$. $\qquad\square$

(4.5) Theorem. *Assume Hypothesis (4.2). Let $\mu_j = \sum_{0 \le i < w_1} \mu_{ij}$, where the characters μ_{ij} are defined in Theorem (4.3), for $0 \le j < w_2$.*

(a) If $\chi_j = \mathrm{Res}^L_K \mu_{ij}$, then χ_j does not depend on i, $\chi_j \in \mathrm{Irr}(K)$ and $\mathrm{Ind}^L_K \chi_j = \mu_j$.

(b) If $\chi \in \mathrm{Irr}(K)$ is not one of the characters χ_j, then $\mathrm{Ind}^L_K \chi \in \mathrm{Irr}(L)$. Moreover, $\mathrm{Ind}^L_K \chi$ is not one of the characters μ_{ij}, and every irreducible character of L is of the form μ_{ij} or $\mathrm{Ind}^L_K \chi$ for some such character χ.

Proof. (a) By (4.3.b), the functions $\mu_{ij} - \mu_{0j} = \delta_j \mathrm{Ind}^L_W(\omega_{ij} - \omega_{0j})$ vanish outside of $(W - W_2)^L$, and so vanish on K since $K \cap W = W_2$. Thus χ_j does not depend on i. Let χ be an irreducible component of χ_j. For all i, $(\mathrm{Ind}^L_K \chi, \mu_{ij}) = (\chi, \mathrm{Res}^L_K \mu_{ij}) \ne 0$, and so

$$w_1 \chi(1) = (\mathrm{Ind}^L_K \chi)(1) \ge \sum_{0 \le i < w_1} \mu_{ij}(1) = w_1 \chi_j(1) \ge w_1 \chi(1).$$

We thus have equality and so $\chi = \chi_j$ and $\mathrm{Ind}^L_K \chi = \mu_j$.

(b) Let $g \in W_1^{\#}$ and let C be a conjugacy class of K normalized by g. If $\langle g \rangle$ acts fixed-point-freely on C, then $|\langle g \rangle|$ divides $|C|$ and so $|K|$; this is a contradiction since $(|K|, w_1) = 1$. There is thus an element $x \in \langle g \rangle^{\#}$ such that $C \cap C_K(x) = C \cap W_2 \ne \emptyset$. The number of conjugacy classes of K normalized by g is thus at most w_2. By [Is], Theorem 6.32, it follows that $\mathrm{Irr}(K)$ has at most w_2 elements left fixed by g. But, since $\mu_{0j} \in \mathrm{Irr}(L)$, χ_j is fixed by g. It follows that, if $\chi \in \mathrm{Irr}(K)$ is not one of the characters χ_j, then χ is not fixed by g. We then see that $I_L(\chi) = K$, and, by (1.5.b), $\mathrm{Ind}^L_K \chi$ is irreducible. Moreover, $(\mathrm{Ind}^L_K \chi, \mu_{ij}) = (\chi, \chi_j) = 0$. Finally, if $\mu \in \mathrm{Irr}(L)$ and if χ is an irreducible component of $\mathrm{Res}^L_K \mu$, then μ is a component of $\mathrm{Ind}^L_K \chi$, and so is one of the characters μ_{ij} or is of the form $\mathrm{Ind}^L_K \chi$. $\qquad\square$

(4.6) Hypothesis. (a) *Let G be a finite group and L a subgroup of G which satisfies Hypothesis (4.2).*

(b) *Assume that G and W satisfy Hypothesis (3.1).*

(c) *Let H be a normal subgroup of L such that $W_2 \subset H \subset K$.*

(d) *Assume that Hypothesis (2.2) holds for a subset A satisfying*

$$\bigcup_{h \in H^\#} C_K(h)^\# \subset A \subset K^\#.$$

Set $A_0 = A \cup V^L$. Assume that Hypothesis (2.2) holds with A_0 in place of A.

(e) *The symbols σ, ω_{ij}, μ_{ij}, μ_j, χ_j and δ_j have the same meaning as in (3.3) and in Theorems (3.2), (4.3) and (4.5), and τ is the Dade isometry relative to A_0.*

(4.7) *Assume Hypothesis (4.6). Let $\chi \in \mathrm{Irr}(K)$ be such that $H \not\subset \mathrm{Ker}\,\chi$. Then $\mathrm{Supp}\,\chi \subset A \cup \{1\}$ and $\mathrm{Supp}\,\mathrm{Ind}_K^L \chi \subset A \cup \{1\}$. If $j \geq 1$, then $H \not\subset \mathrm{Ker}\,\chi_j$, $\mathrm{Supp}\,\chi_j \subset A \cup \{1\}$ and $\mathrm{Supp}\,\mu_j \subset A \cup \{1\}$.*

Proof. By (4.6.d) and (1.2), $\mathrm{Supp}\,\chi \subset A \cup \{1\}$. Thus, $\mathrm{Supp}\,\mathrm{Ind}_K^L \chi \subset A \cup \{1\}$. Then it suffices to show that $H \not\subset \mathrm{Ker}\,\chi_j$ for $j \geq 1$. Suppose that $H \subset \mathrm{Ker}\,\chi_j$. As $W_2 \subset H$,

$$\omega_{0j}(y) = \omega_{0j}(xy) = \delta_j \mu_{0j}(xy) = \delta_j \mu_{0j}(x) = \omega_{0j}(x) = 1$$

for $x \in W_1^\#$ and $y \in W_2$ by Theorem (4.3). Thus $j = 0$, contrary to the hypothesis. $\qquad\square$

(4.8) *Assume Hypothesis (4.6). Let i, j and k be indices such that $0 \leq i < w_1$, $0 < j, k < w_2$ and $\mu_{ij}(1) = \mu_{ik}(1)$. Then*

$$\mathrm{Supp}(\mu_{ij} - \mu_{ik}) \subset A_0, \quad \delta_j = \delta_k \quad \text{and} \quad (\mu_{ij} - \mu_{ik})^\tau = \delta_j(\omega_{ij}^\sigma - \omega_{ik}^\sigma).$$

Proof. By (4.3.d), $\delta_j \equiv \delta_k \pmod{w_1}$, and so $\delta_j = \delta_k$ because $w_1 > 2$. By (4.3.c), $\mu_{ij} - \mu_{ik}$ vanishes on W_1. If $z \in L - K$, then, by (2.1), z is conjugate in L to an element of $x C_K(x) = x W_2$ for some $x \in W_1^\#$. Thus $z \in V^L$ if $z \in \mathrm{Supp}(\mu_{ij} - \mu_{ik})$. On the other hand, $\mathrm{Supp}(\mu_{ij} - \mu_{ik}) \cap K \subset A$ by (4.7). Thus, $\mathrm{Supp}(\mu_{ij} - \mu_{ik}) \subset A_0$. By (3.2.c), (4.3.c) and the definition of τ, $\psi = (\mu_{ij} - \mu_{ik})^\tau - \delta_j(\omega_{ij}^\sigma - \omega_{ik}^\sigma)$ vanishes on V. Since τ is an isometry and $(\mu_{ij} - \mu_{ik})^\tau$ vanishes on 1, there are characters $\lambda_1, \lambda_2 \in \mathrm{Irr}(G)$ such that $\psi = \lambda_1 - \lambda_2 - \delta_j(\omega_{ij}^\sigma - \omega_{ik}^\sigma)$. Using the notation of Hypothesis (3.6), we see that $NC(\psi) \leq 4 < 2\inf(w_1, w_2)$. Cases (b) and (c) of (3.8) are impossible for ψ because, in these cases, ψ contains at least three components of the form ω_{rs}^σ with the same coefficient. By (3.8), ψ is thus orthogonal to ω_{ij}^σ and ω_{ik}^σ, which implies that $\psi = 0$. $\qquad\square$

(4.9) Theorem. *Assume Hypothesis* (4.6). *Let k be such that $0 < k < w_2$ and let $\mathcal{T} = \{\mu_j \mid 0 < j < w_2,\ \mu_j(1) = \mu_k(1)\}$.*

(a) *If $\mu_j \in \mathcal{T}$, then $\overline{\mu_j} \in \mathcal{T}$ and $\overline{\mu_j} \neq \mu_j$. Also,*

$$0 \neq \mathbf{Z}[\mathcal{T}, L^{\#}] = \mathbf{Z}[\mathcal{T}, A].$$

(b) *The $\mathbf{Z}$-linear mapping from $\mathbf{Z}[\mathcal{T}]$ to $\mathbf{Z}[\mathrm{Irr}\,G]$ which sends the character μ_j to $\delta_k \sum_{0 \leq i < w_1} \omega_{ij}^{\sigma}$ is an isometry which coincides with τ on $\mathbf{Z}[\mathcal{T}, A]$.*

Proof. (a) For all i such that $0 \leq i < w_1$, let $\overline{\omega_{ij}} = \omega_{i'j'}$. Then $j' \neq j$ because $j \neq 0$ and $|W|$ is odd, and j' does not depend on i. It follows from (3.9) and Theorem (4.3) that $\delta_j \overline{\mu_{ij}} = \delta_{j'} \mu_{i'j'}$, and so $\overline{\mu_{ij}} = \mu_{i'j'}$. Thus $\overline{\mu_j} = \mu_{j'} \neq \mu_j$. In particular, $0 \neq \overline{\mu_k} - \mu_k \in \mathbf{Z}[\mathcal{T}, L^{\#}]$ and, by (4.7), $\mathbf{Z}[\mathcal{T}, L^{\#}] = \mathbf{Z}[\mathcal{T}, A]$.

(b) It is clear that the mapping defined in (b) is an isometry. Furthermore, the $\mathbf{Z}$-module $\mathbf{Z}[\mathcal{T}, A]$ is generated by the functions $\mu_j - \mu_k$ for $\mu_j \in \mathcal{T}$. But, by (4.8),

$$(\mu_j - \mu_k)^{\tau} = \sum_{0 \leq i < w_1} (\mu_{ij} - \mu_{ik})^{\tau} = \delta_k \sum_{0 \leq i < w_1} (\omega_{ij}^{\sigma} - \omega_{ik}^{\sigma})$$

which proves (b). $\square$

(4.10) *Assume Hypothesis* (4.6). *For $0 \leq i < w_1$ and $0 \leq j < w_2$,*

$$(\delta_j \mu_{ij} - \delta_j \mu_{0j} - \mu_{i0} + \mu_{00})^{\tau} = \omega_{ij}^{\sigma} - \omega_{0j}^{\sigma} - \omega_{i0}^{\sigma} + \omega_{00}^{\sigma}.$$

Proof. Let $\alpha = \omega_{ij} - \omega_{0j} - \omega_{i0} + \omega_{00}$ and $\beta = \delta_j \mu_{ij} - \delta_j \mu_{0j} - \mu_{i0} + \mu_{00}$. By (4.3.b) and (4.4),

$$\beta = \mathrm{Ind}_W^L(\omega_{ij} - \omega_{0j}) - \mathrm{Ind}_W^L(\omega_{i0} - \omega_{00}) = \mathrm{Ind}_W^L \alpha.$$

By (3.4), $\mathrm{Supp}(\alpha) \subset V$ and so $\mathrm{Supp}(\beta) \subset V^L$. For $x \in V$, $C_G(x) = W \subset L$. By definition of τ, it follows that

$$\beta^{\tau}(g) = \begin{cases} \beta(g) & \text{if } g \in V, \\ 0 & \text{if } g \notin V^G. \end{cases}$$

Thus, by (4.3.c) and (3.2.c), $\beta^{\tau}(g) = \alpha^{\sigma}(g)$ for all $g \in G$. $\square$

5. Coherence

(5.1) Definition. Let L and G be finite groups, $A \subset L$ and $\mathcal{S} \subset \mathbf{Z}[\mathrm{Irr}\,L]$. Let τ be a $\mathbf{Z}$-linear isometry from E to $\mathbf{Z}[\mathrm{Irr}\,G]$, where E is a $\mathbf{Z}$-module such that $\mathbf{Z}[\mathcal{S}, A] \subset E \subset \mathbf{Z}[\mathrm{Irr}\,L]$. We say that $(\mathcal{S}, A, \tau)$ is *coherent*, or that $\mathcal{S}$ is *coherent*, if $\mathbf{Z}[\mathcal{S}, A] \neq 0$ and if there is a linear isometry from $\mathbf{Z}[\mathcal{S}]$ to $\mathbf{Z}[\mathrm{Irr}\,G]$ which coincides with τ on $\mathbf{Z}[\mathcal{S}, A]$.

(5.2) Hypothesis. (a) *Let L and G be finite groups and let $\mathcal{S}$ be a non-empty set of characters of L. Assume that, if $\chi \in \mathcal{S}$, then $\overline{\chi} \in \mathcal{S}$ and $\overline{\chi} \neq \chi$.*

(b) *Assume that τ is a linear isometry from $\mathbf{Z}[\mathcal{S}, L^{\#}]$ to $\mathbf{Z}[\mathrm{Irr}\,G, G^{\#}]$.*

(c) *The elements of $\mathcal{S}$ are pairwise orthogonal.*

(d) *Assume that, for $\chi \in \mathcal{S}$, $(\chi - \overline{\chi})^{\tau} = \sum_{\alpha \in R(\chi)} \alpha$ for some orthonormal subset $R(\chi)$ of $\mathbf{Z}[\mathrm{Irr}\,G]$.*

(e) *If $\chi \in \mathcal{S}$, $\phi \in \mathcal{S}$ and ϕ is orthogonal to $\{\chi, \overline{\chi}\}$, then $R(\phi)$ is orthogonal to $R(\chi)$.*

(5.3) (a) *Assume (5.2.a), (5.2.b) and that $\mathcal{S} \subset \mathrm{Irr}\,L$. Then Hypothesis (5.2) holds.*

(b) *Assume Hypothesis (4.6), (5.2.a) and that*

$$\mathcal{S} \subset \{\mathrm{Ind}_K^L \theta \mid \theta \in \mathrm{Irr}\,K, \ H \not\subset \mathrm{Ker}\,\theta\}.$$

Then Hypothesis (5.2) holds with the isometry τ of Hypothesis (5.2) being the restriction to $\mathbf{Z}[\mathcal{S}, L^{\#}]$ of the isometry τ of Hypothesis (4.6). If $\phi \in \mathcal{S} \cap \mathrm{Irr}\,L$, then $R(\phi)$ is orthogonal to ω^{σ} for all $\omega \in \mathrm{Irr}(W)$.

Proof. (a) Hypothesis (5.2.c) is clear. For $\chi \in \mathcal{S}$, $\|(\chi - \overline{\chi})^{\tau}\|^2 = 2$ and so (5.2.d) holds with $|R(\chi)| = 2$. If χ, $\phi \in \mathcal{S}$ and ϕ is orthogonal to $\{\chi, \overline{\chi}\}$, then we have $((\phi - \overline{\phi})^{\tau}, (\chi - \overline{\chi})^{\tau}) = 0$ and (5.2.e) follows from (4.1).

(b) By (4.7), $\mathbf{Z}[\mathcal{S}, L^{\#}] = \mathbf{Z}[\mathcal{S}, A]$, and so τ is defined on $\mathbf{Z}[\mathcal{S}, L^{\#}]$. The elements of $\mathcal{S}$ are pairwise orthogonal by (1.5.c). Property (5.2.d) holds if χ is irreducible, as in (a). Otherwise, by (4.4) and Theorem (4.5), χ is of the form μ_j, $0 < j < w_2$. By Theorem (4.9), (5.2.d) holds for μ_j with

$$R(\mu_j) = \{\delta_j \omega_{ij}^{\sigma}, -\delta_j \omega_{ik}^{\sigma} \mid 0 \leq i < w_1\},$$

where k satisfies $\overline{\mu_j} = \mu_k$. We show that (5.2.e) holds. If χ and ϕ are irreducible, (5.2.e) follows from (4.1). If χ and ϕ are reducible, (5.2.e) follows from the form of $R(\mu_j)$. Suppose that $\phi \in \mathcal{S} \cap \mathrm{Irr}(L)$. By (4.7), $\mathrm{Supp}(\phi - \overline{\phi}) \subset A$. By the definition of τ, $(\phi - \overline{\phi})^{\tau}$ vanishes on V. In the notation of Hypothesis (3.6), $NC((\phi - \overline{\phi})^{\tau}) \leq \|\phi - \overline{\phi}\|^2 = 2$, and so, by (3.8), $R(\phi)$ is orthogonal to ω^{σ} for all $\omega \in \mathrm{Irr}\,W$ and, in particular, to $R(\mu_j)$ if $\mu_j \in \mathcal{S}$. $\qquad\square$

(5.4) *Assume Hypothesis (5.2). Let $\chi \in \mathcal{S}$ and $\psi \in \mathbf{Z}[\mathcal{S}]$ be such that*

$$(\chi, \psi) = (\overline{\chi}, \psi) = 0.$$

Let τ_1 be an isometry from $\mathbf{Z}[\chi - \psi, \chi - \overline{\chi}]$ to $\mathbf{Z}[\mathrm{Irr}\, G]$ which coincides with τ on $\mathbf{Z}[\chi - \overline{\chi}]$. Set $(\chi - \psi)^{\tau_1} = X - Y$, where $X \in \mathbf{Z}[R(\chi)]$ and Y is orthogonal to $R(\chi)$.

   (a) $\|X\|^2 \geq \|\chi\|^2$.

   (b) *Suppose that $\|Y\|^2 \geq \|\psi\|^2$. Then $\|X\|^2 = \|\chi\|^2$, $\|Y\|^2 = \|\psi\|^2$ and $X = \sum_{\alpha \in E} \alpha$ for some subset $E \subset R(\chi)$.*

Proof. (a) We note first that

$$\|\chi\|^2 = (\chi - \psi, \chi - \overline{\chi}) = \left(X - Y, \sum_{\alpha \in R(\chi)} \alpha \right) = \sum_{\alpha \in R(\chi)} (X, \alpha).$$

For $\alpha \in R(\chi)$, $(X, \alpha) \in \mathbf{Z}$ and so $(X, \alpha) \leq (X, \alpha)^2$. Thus,

$$\|\chi\|^2 \leq \sum_{\alpha \in R(\chi)} (X, \alpha)^2 = \|X\|^2.$$

   (b) By (a),

$$\|\chi\|^2 + \|\psi\|^2 = \|\chi - \psi\|^2 = \|X\|^2 + \|Y\|^2 \geq \|\chi\|^2 + \|\psi\|^2.$$

We then conclude that $\|X\|^2 = \|\chi\|^2$, that $\|Y\|^2 = \|\psi\|^2$ and that the inequalities in the proof of (a) are equalities. Thus, for all $\alpha \in R(\chi)$, $(X, \alpha)^2 = (X, \alpha)$ and $(X, \alpha) = 0$ or 1. $\square$

(5.5) *Assume Hypothesis (5.2). Let $\chi \in \mathcal{S}$ and let τ_1 be an isometry from $\mathbf{Z}[\chi, \overline{\chi}]$ to $\mathbf{Z}[\mathrm{Irr}\, G]$ which coincides with τ on $\mathbf{Z}[\chi - \overline{\chi}]$. Then $\chi^{\tau_1} = \sum_{\alpha \in E} \alpha$ for some subset $E \subset R(\chi)$.*

Proof. We apply (5.4) with $\psi = 0$. The hypothesis of (5.4.b) is satisfied, and so $Y = 0$ and $\chi^{\tau_1} = X = \sum_{\alpha \in E} \alpha$. $\square$

(5.6) Theorem. *Assume Hypothesis (5.2). Let $\mathcal{S}_1 = \{\chi_1, \ldots, \chi_n\}$ be a subset of $\mathcal{S}$ closed under complex conjugation, where $|\mathcal{S}_1| = n$, and let $\mathcal{S}_2 = \{\chi, \overline{\chi}\}$ be a subset of $\mathcal{S}$ such that $\mathcal{S}_1 \cap \mathcal{S}_2 = \emptyset$. Assume that*

   (a) $\mathcal{S}_1$ *is coherent,*

   (b) $\chi_1(1)$ *divides* $\chi(1)$,

   (c) $2\chi(1)\chi_1(1) < \displaystyle\sum_{i=1}^{n} \frac{\chi_i(1)^2}{\|\chi_i\|^2}.$

Then $\mathcal{S}_1 \cup \mathcal{S}_2$ is coherent.

Proof. Set $\chi(1) = a\chi_1(1)$ and $\chi_i(1) = a_i\chi_1(1)$ for $1 \leq i \leq n$. Let τ_1 be an isometry from $\mathbf{Z}[\mathcal{S}_1]$ to $\mathbf{Z}[\mathrm{Irr}\,G]$ which extends the restriction of τ to $\mathbf{Z}[\mathcal{S}_1, L^\#]$. Set $(\chi_i - a_i\chi_1)^\tau = \dfrac{1}{\chi_1(1)}(\chi_1(1)\chi_i - \chi_i(1)\chi_1)^\tau$; this is compatible with previous notation if $a_i \in \mathbf{N}$.

(5.6.1) *Let* $(\chi - a\chi_1)^\tau = X - Y$, *where* $X \in \mathbf{Z}[R(\chi)]$ *and* Y *is orthogonal to* $R(\chi)$. *There is an integer* $\lambda \in \mathbf{Z}$ *such that*

$$Y = a\chi_1^{\tau_1} - \lambda \sum_{i=1}^{n} \frac{a_i}{\|\chi_i\|^2}\chi_i^{\tau_1} + Z,$$

where $Z \in \mathrm{CF}(G)$ *is orthogonal to* $\mathcal{S}_1^{\tau_1}$.

Proof. Set $Y = a\chi_1^{\tau_1} - \sum_{i=1}^{n} \lambda_i \chi_i^{\tau_1} + Z$ with $\lambda_i \in \mathbf{C}$ and where $Z \in \mathrm{CF}(G)$ is orthogonal to $\mathcal{S}_1^{\tau_1}$. For $1 \leq i \leq n$, $\chi_i^{\tau_1}$ is orthogonal to $R(\chi)$ by (5.5) and (5.2.e). It follows that, for $1 < i \leq n$,

$$
\begin{aligned}
aa_i\|\chi_1\|^2 &= ((\chi - a\chi_1)^\tau, (\chi_i - a_i\chi_1)^\tau) = (X - Y, \chi_i^{\tau_1} - a_i\chi_1^{\tau_1}) \\
&= (-Y, \chi_i^{\tau_1} - a_i\chi_1^{\tau_1}) = \lambda_i\|\chi_i\|^2 + a_i(a - \lambda_1)\|\chi_1\|^2.
\end{aligned}
$$

Thus, $\lambda_i = \dfrac{a_i\lambda_1\|\chi_1\|^2}{\|\chi_i\|^2} = \lambda\dfrac{a_i}{\|\chi_i\|^2}$, where $\lambda = \lambda_1\|\chi_1\|^2$. Also, $\lambda_1 = \lambda\dfrac{a_1}{\|\chi_1\|^2}$ since $a_1 = 1$. As $(\chi - a\chi_1)^\tau \in \mathbf{Z}[\mathrm{Irr}\,G]$,

$$(Y, \chi_1^{\tau_1}) = \left(a - \frac{\lambda}{\|\chi_1\|^2}\right)\|\chi_1\|^2 \in \mathbf{Z},$$

and so $\lambda \in \mathbf{Z}$. $\qquad\square$

(5.6.2) $Y = a\chi_1^{\tau_1}$.

Proof. We note first that $\|\chi\|^2 + a^2\|\chi_1\|^2 = \|(\chi - a\chi_1)^\tau\|^2 = \|X\|^2 + \|Y\|^2$. By (5.4.a), $\|X\|^2 \geq \|\chi\|^2$, and so $\|Y\|^2 \leq a^2\|\chi_1\|^2$. Then, by (5.6.1),

$$\left(\frac{\lambda}{\|\chi_1\|^2} - a\right)^2\|\chi_1\|^2 + \lambda^2\sum_{i=2}^{n}\frac{a_i^2}{\|\chi_i\|^4}\|\chi_i\|^2 + \|Z\|^2 \leq a^2\|\chi_1\|^2,$$

or

$$\lambda^2\sum_{i=1}^{n}\frac{a_i^2}{\|\chi_i\|^2} - 2\lambda a + \|Z\|^2 \leq 0.$$

Let $b = \dfrac{2a}{\displaystyle\sum_{i=1}^{n}\frac{a_i^2}{\|\chi_i\|^2}}$. By hypothesis (c), $2a\chi_1(1)^2 < \sum \dfrac{a_i^2\chi_1(1)^2}{\|\chi_i\|^2}$, whence it follows that $0 < b < 1$. Since $\lambda \in \mathbf{Z}$ and since $\lambda^2 - b\lambda \leq 0$, it follows that $\lambda = 0$ and that $Z = 0$. $\qquad\square$

(5.6.3) $\mathcal{S}_1 \cup \mathcal{S}_2$ *is coherent.*

Proof. Let τ_2 be the $\mathbf{Z}$-linear mapping from $\mathbf{Z}[\mathcal{S}_1 \cup \mathcal{S}_2]$ to $\mathbf{Z}[\operatorname{Irr} G]$ which extends τ_1 and satisfies $\chi^{\tau_2} = X$ and $\overline{\chi}^{\,\tau_2} = X - (\chi - \overline{\chi})^\tau$. Then τ_2 coincides with τ on $\mathbf{Z}[\mathcal{S}_1, L^{\#}]$, on $\chi - a\chi_1$ and on $\chi - \overline{\chi}$, which generate $\mathbf{Z}[\mathcal{S}_1 \cup \mathcal{S}_2, L^{\#}]$. By (5.5) and (5.2.e), χ^{τ_2} and $\overline{\chi}^{\tau_2}$ are orthogonal to $\mathcal{S}_1^{\tau_2}$. As $\|Y\|^2 = \|a\chi_1\|^2$, (5.4.b) shows that $X = \sum_{\alpha \in E} \alpha$ for some subset $E \subset R(\chi)$ such that $|E| = \|\chi\|^2$. It follows that $\overline{\chi}^{\tau_2} = -\sum_{\alpha \in R(\chi) - E} \alpha$, and so

$$\|\overline{\chi}^{\tau_2}\|^2 = |R(\chi) - E| = |R(\chi)| - |E| = \|\chi - \overline{\chi}\|^2 - \|\chi\|^2 = \|\overline{\chi}\|^2$$

and $(\chi^{\tau_2}, \overline{\chi}^{\tau_2}) = 0$. $\qquad\qquad\Box$

(5.7) *Assume Hypothesis (5.2) and that $\chi(1)$ is independent of χ for $\chi \in \mathcal{S}$. Then $\mathcal{S}$ is coherent.*

Proof. If $|\mathcal{S}| = 2$, this follows from (5.2.d). Suppose that $\mathcal{S} = \mathcal{S}_1 \cup \{\chi, \overline{\chi}\}$ where $\mathcal{S}_1 \neq \emptyset$ is orthogonal to $\{\chi, \overline{\chi}\}$. Let $\chi_1 \in \mathcal{S}_1$. By (5.2.e), $R(\chi)$ is orthogonal to $R(\chi_1)$. Let $(\chi - \chi_1)^\tau = X - X_1 + Y$, where $X \in \mathbf{Z}[R(\chi)]$, $X_1 \in \mathbf{Z}[R(\chi_1)]$ and Y is orthogonal to $R(\chi)$ and to $R(\chi_1)$. From (5.4.a) we obtain $\|X\|^2 \geq \|\chi\|^2$ and $\|X_1\|^2 \geq \|\chi_1\|^2$, and so $\|X_1 - Y\|^2 \geq \|\chi_1\|^2$. By (5.4.b), it follows that $\|X\|^2 = \|\chi\|^2$, that $\|X_1 - Y\|^2 = \|\chi_1\|^2$ whence $Y = 0$, and that $X = \sum_{\alpha \in E} \alpha$ for some subset $E \subset R(\chi)$. We show that X is independent of $\chi_1 \in \mathcal{S}_1$. We know that

$$(\chi - \overline{\chi_1})^\tau = X - X_1 + (\chi_1 - \overline{\chi_1})^\tau = X - X_1'$$

for some $X_1' \in \mathbf{Z}[R(\chi_1)]$. If $\chi_2 \in \mathcal{S}_1 - \{\chi_1, \overline{\chi_1}\}$, let $(\chi - \chi_2)^\tau = X' - X_2$, where $X' = \sum_{\alpha \in E'} \alpha$ for some subset $E' \subset R(\chi)$ and $X_2 \in \mathbf{Z}[R(\chi_2)]$. Consequently,

$$|E| = \|\chi\|^2 = (\chi - \chi_1, \chi - \chi_2) = (X - X_1, X' - X_2) = (X, X') = |E \cap E'|.$$

Thus, $E = E'$ and $X = X'$. Let τ_1 be the $\mathbf{Z}$-linear mapping from $\mathbf{Z}[\mathcal{S}]$ to $\mathbf{Z}[\operatorname{Irr} G]$ for which $\chi^{\tau_1} = X$ and $\chi_1^{\tau_1} = X - (\chi - \chi_1)^\tau$ for $\chi_1 \in \mathcal{S}_1$ or $\chi_1 = \overline{\chi}$. Since $\|X\|^2 = \|\chi\|^2$, since $(X, (\chi - \chi_1)^\tau) = \|\chi\|^2$ for $\chi_1 \in \mathcal{S}_1$ or $\chi_1 = \overline{\chi}$ and since $\mathbf{Z}[\mathcal{S}]$ is generated by χ and the functions $\chi - \chi_1$, it follows that τ_1 is an isometry. $\qquad\Box$

(5.8) *Suppose that the hypothesis of (5.3.b) holds, that $\mathcal{S} \cap \operatorname{Irr}(L) \neq \emptyset$ and that $\mu_k \in \mathcal{S}$ for some $k \geq 1$. Let $\mu_j = \overline{\mu_k}$, and let τ_1 be an isometry from $\mathbf{Z}[\mathcal{S}]$ to $\mathbf{Z}[\operatorname{Irr} G]$ which coincides with τ on $\mathbf{Z}[\mathcal{S}, L^{\#}]$. Then $\mu_k^{\tau_1} = \delta_k \sum_{0 \leq i < w_1} \omega_{ik}^\sigma$ or $\mu_k^{\tau_1} = -\delta_k \sum_{0 \leq i < w_1} \omega_{ij}^\sigma$. In the second case, j and k are the only indices ℓ such that $\ell \geq 1$, $\mu_\ell \in \mathcal{S}$ and $\mu_\ell(1) = \mu_k(1)$.*

Proof. By (5.5), $\mu_k^{\tau_1} = \sum_i a_{ik} \omega_{ik}^\sigma + \sum_i a_{ij} \omega_{ij}^\sigma$, where $a_{ik} \in \{0, \delta_k\}$ and where $a_{ij} \in \{0, -\delta_k\}$. Let $\chi \in \mathcal{S} \cap \operatorname{Irr}(L)$. By (5.3.b) and (5.5), χ^{τ_1} is orthogonal to the image of σ, and so vanishes on V by (3.2.d). On the other hand, by (4.7), $\chi(1)\mu_k - \mu_k(1)\chi \in \mathbf{Z}[\mathcal{S}, A]$. As $A \cap V = \emptyset$, it follows from the definition of τ that $(\chi(1)\mu_k - \mu_k(1)\chi)^\tau$ vanishes on V. Thus $\mu_k^{\tau_1}$ vanishes on V. By (3.7),

$$a_{ik} = a_{i0} + a_{0k} - a_{00} = a_{0k} \quad \text{and} \quad a_{ij} = a_{i0} + a_{0j} - a_{00} = a_{0j}$$

for all i. As $\|\mu_k^{\tau_1}\|^2 = w_1$, it follows that $\mu_k^{\tau_1} = \delta_k \sum_i \omega_{ik}^\sigma$ or $\mu_k^{\tau_1} = -\delta_k \sum_i \omega_{ij}^\sigma$. Suppose that there is an index $\ell \geq 1$ such that $\ell \neq j$, $\ell \neq k$, $\mu_\ell \in \mathcal{S}$ and $\mu_\ell(1) = \mu_k(1)$. By Theorem (4.9), $\mu_k^{\tau_1} - \mu_\ell^{\tau_1} = \delta_k \sum_i (\omega_{ik}^\sigma - \omega_{i\ell}^\sigma)$. Since $\mu_\ell^{\tau_1}$ is a sum of elements of $R(\mu_\ell)$, it follows that $\mu_k^{\tau_1} = \delta_k \sum_i \omega_{ik}^\sigma$. $\qquad\square$

(5.9) *Assume Hypothesis (2.2).*

(a) *Let $\mathcal{S} \subset \operatorname{Irr} L$ be such that $\mathbf{Z}[\mathcal{S}, L^\#] = \mathbf{Z}[\mathcal{S}, A]$ and $|\mathcal{S}| \geq 2$. Let u be an automorphism of the field $\mathbf{Q}_{|G|}$ for which $\mathcal{S}^u \subset \mathcal{S}$. Let τ_1 be a linear isometry from $\mathbf{Z}[\mathcal{S}]$ to $\mathbf{Z}[\operatorname{Irr} G]$ which coincides with τ on $\mathbf{Z}[\mathcal{S}, A]$. Then, if $\chi \in \mathcal{S}$, $\chi^{\tau_1 u} = \chi^{u\tau_1}$.*

(b) *Let $\chi \in \operatorname{Irr} L$ be such that $\operatorname{Supp}(\chi) \subset A \cup \{1\}$. Then there is a character $\mu \in \operatorname{Irr} G$ such that $(\chi - \overline{\chi})^\tau = \mu - \overline{\mu}$.*

Proof. (a) Let $\psi \in \mathcal{S}$, $\psi \neq \chi$. By the definition of τ,

$$
\begin{aligned}
\psi(1)\chi^{\tau_1 u} &- \chi(1)\psi^{\tau_1 u} \\
&= (\psi(1)\chi - \chi(1)\psi)^{\tau u} = (\psi(1)\chi - \chi(1)\psi)^{u\tau} = \psi(1)\chi^{u\tau_1} - \chi(1)\psi^{u\tau_1}.
\end{aligned}
$$

For all $\phi \in \mathcal{S}$, $\|\phi^{\tau_1}\| = 1$ and $\chi(1)\phi^{\tau_1} - \phi(1)\chi^{\tau_1} = (\chi(1)\phi - \phi(1)\chi)^\tau$ vanishes on 1. There is thus an integer $\varepsilon = \pm 1$ such that $\varepsilon \phi^{\tau_1} \in \operatorname{Irr} G$ for all $\phi \in \mathcal{S}$. The equation then implies that $\chi^{\tau_1 u} = \chi^{u\tau_1}$.

(b) We may assume that $\overline{\chi} \neq \chi$. Since $\|\chi - \overline{\chi}\|^2 = 2$ and $(\chi - \overline{\chi})^\tau(1) = 0$, there are characters μ, $\mu' \in \operatorname{Irr} G$ such that $(\chi - \overline{\chi})^\tau = \mu - \mu'$. Therefore, by (a), $\mu' = \overline{\mu}$. $\qquad\square$

6. Some Coherence Theorems

(6.1) Hypothesis. *Assume that Hypothesis (5.2) holds. We assume that K is a solvable normal subgroup of L and that $\mathcal{S} = \{\mathrm{Ind}_K^L \theta \mid \theta \in \mathrm{Irr}\, K,\ \theta \neq 1_K\}$. If A is a normal subgroup of L contained in K, set*

$$\mathcal{S}(A) = \{\mathrm{Ind}_K^L \theta \mid \theta \in \mathrm{Irr}\, K,\ A \subset \mathrm{Ker}\,\theta,\ \theta \neq 1_K\}.$$

(6.2) *Assume Hypothesis (6.1). Let A, B, C and D be normal subgroups of L such that*

(a) $A \subsetneq K$, $B \subset D \subset C \subset K$ *and* $D/B \subset Z(C/B)$,

(b) $\mathcal{S}(A)$ *is coherent but* $\mathcal{S}(B)$ *is not.*

Then $2|L : C|\sqrt{|C : D|} \geq |K : A| - 1$.

Proof. By (6.2.b), there are sets $\mathcal{S}_1$ and $\mathcal{S}_2$, with $\mathcal{S}_1$ closed under complex conjugation, such that $\mathcal{S}(A) \subset \mathcal{S}_1 \subset \mathcal{S}(A) \cup \mathcal{S}(B)$ and $\mathcal{S}_2 = \{\psi, \overline{\psi}\} \subset \mathcal{S}(B)$ and such that $\mathcal{S}_1$ is coherent but $\mathcal{S}_1 \cup \mathcal{S}_2$ is not coherent. Since K is solvable, K/A has a non-trivial irreducible character of degree 1, whence $\mathcal{S}(A)$ contains a character of degree $|L : K|$, and $|L : K|$ divides $\psi(1)$. By Theorem (5.6),

$$2\psi(1)|L : K| \geq \sum_{\chi \in \mathcal{S}_1} \frac{\chi(1)^2}{\|\chi\|^2} \geq \sum_{\chi \in \mathcal{S}(A)} \frac{\chi(1)^2}{\|\chi\|^2}.$$

Let $T = \{\theta \in \mathrm{Irr}\, K \mid A \subset \mathrm{Ker}\,\theta,\ \theta \neq 1_K\}$. By (1.5.c, d),

$$\sum_{\chi \in \mathcal{S}(A)} \frac{\chi(1)^2}{\|\chi\|^2} = |L : K| \sum_{\theta \in T} \theta(1)^2 = |L : K|(|K : A| - 1).$$

Thus, $2\psi(1) \geq |K : A| - 1$. Let $\theta \in \mathrm{Irr}\, K$ be such that $\psi = \mathrm{Ind}_K^L \theta$. By (1.8), $\theta(1) \leq |K : C|\sqrt{|C : D|}$. Thus, $\psi(1) \leq |L : C|\sqrt{|C : D|}$ and the required conclusion follows. $\qquad\square$

(6.3) Theorem. *Assume Hypothesis (6.1). Let M, H and H_1 be normal subgroups of L such that $M \subset H_1 \subset H \subset K$. Assume further that*

(a) H/M *is nilpotent,*

(b) $\mathcal{S}(H_1)$ *is coherent,*

(c) $|H : H_1| > 4|L : K|^2 + 1$.

Then $\mathcal{S}(M)$ is coherent.

Proof. By (b), there is a normal subgroup A of L such that $M \subset A \subset H_1$ and $\mathcal{S}(A)$ is coherent, and which is minimal with these properties. Suppose that $A \neq M$. Let B be a normal subgroup of L such that $M \subset B \subsetneq A$, and which is maximal with these properties. Since H/M is nilpotent, we have $(A/B) \cap Z(H/B) \neq 1$ and, by the maximality of B, $A/B \subset Z(H/B)$. By

(6.2) with $C = H$ and $D = A$,
$$2|L : H|\sqrt{|H : A|} \geq |K : A| - 1.$$
Writing $x = |H : A|$, we obtain
$$2|L : K||K : H|\sqrt{x} \geq |K : H|x - 1$$
and so
$$2|L : K| \geq \sqrt{x} - \frac{1}{|K : H|\sqrt{x}} \geq \sqrt{x} - \frac{1}{\sqrt{x}}$$
and
$$\left(\sqrt{x} - \frac{1}{\sqrt{x}}\right)^2 = x - 2 + \frac{1}{x} \leq 4|L : K|^2.$$

By (c), $x \geq |H : H_1| > 1$. Then $|H : H_1| - 1 \leq x - 1 \leq 4|L : K|^2$ as $x \in \mathbf{N}$ and $1/x < 1$; this contradicts inequality (c). $\qquad\square$

(6.4) Hypothesis. (a) *Assume that Hypothesis (6.1) holds and that $|L|$ is odd.*

(b) *Let M be a normal subgroup of L contained in K such that K/M is nilpotent.*

(c) *Let H_1/M be the commutator subgroup of K/M. Assume that L/H_1 is a Frobenius group with kernel K/H_1.*

(6.5) *Assume Hypothesis (6.4) and that $S(M)$ is not coherent.*

(a) *K/H_1 is a chief factor of L and $|K : H_1| \leq 4|L : K|^2 + 1$.*

(b) *There is a prime number p such that K/M is a non-abelian p-group.*

(c) *$|L : K|$ does not divide $p - 1$.*

Proof. (a) Hypothesis (a) of Theorem (6.3) holds with $H = K$. Since K/H_1 is abelian and non-trivial, (6.3.b) holds by (5.7). Therefore, from Theorem (6.3), we obtain $|K : H_1| \leq 4|L : K|^2 + 1$. Suppose that there is a normal subgroup H_2 of L such that $H_1 \subsetneqq H_2 \subsetneqq K$. By (6.4.c), $|L : K|$ then divides $|K : H_2| - 1$ and $|H_2 : H_1| - 1$. As $|L|$ is odd, it follows that
$$|K : H_1| = |K : H_2||H_2 : H_1| \geq (2|L : K| + 1)^2 > 4|L : K|^2 + 1,$$
which is a contradiction.

(b) Since $S(H_1)$ is coherent, $M \neq H_1$, and so K/M is not abelian. Since K/M is a nilpotent group whose commutator subgroup is H_1/M and since K/H_1 is a chief factor of L, K/M is a p-group for some prime number p.

(c) If $|L : K|$ divides $p - 1$, then $p \geq 2|L : K| + 1$. Since K/M is a non-abelian p-group, $|K : H_1| \geq p^2 \geq (2|L : K| + 1)^2 > 4|L : K|^2 + 1$, which is a contradiction. $\qquad\square$

(6.6) *Suppose that Hypothesis (6.4) holds with $M = 1$. Let Z be a normal subgroup of L such that $1 \neq Z \subset Z(K)$ and let $\mathcal{X} = S - S(Z)$. Suppose that $\mathcal{X} \subset \operatorname{Irr} L$. Then $\mathcal{X} = \{\chi \in \operatorname{Irr} L \mid Z \not\subset \operatorname{Ker} \chi\}$ and $\mathcal{X}$ is coherent.*

Proof. Let $n = |\mathcal{X}|$. By (1.1), $n \geq 2$ and so $\mathbf{Z}[\mathcal{X}, L^{\#}] \neq 0$. Let $\chi \in \operatorname{Irr} L$ be such that $Z \not\subset \operatorname{Ker} \chi$. There is a character $\theta \in \operatorname{Irr} K$ for which χ is an irreducible component of $\operatorname{Ind}_K^L \theta$; by (1.6), $Z \not\subset \operatorname{Ker} \theta$. Since $\mathcal{X} \subset \operatorname{Irr} L$, $\chi = \operatorname{Ind}_K^L \theta$. Thus, $\mathcal{X} = \{\chi \in \operatorname{Irr} L \mid Z \not\subset \operatorname{Ker} \chi\}$. We show that $\mathcal{X}$ is coherent. By (6.5), we may assume that K is a p-group for some prime number p. Set $\mathcal{X} = \{\chi_1, \ldots, \chi_n\}$, where $\chi_1(1) \leq \cdots \leq \chi_n(1)$ and $\chi_i = \operatorname{Ind}_K^L \theta_i$, $\theta_i \in \operatorname{Irr} K$. Let k be maximal such that $\chi_k(1) = \chi_1(1)$. By (1.1) and (1.4), $\{\chi_1, \ldots, \chi_k\}$ is coherent. Let i be such that $k < i \leq n$. Then

$$\sum_{1 \leq j < i} \chi_j(1)^2 = |L| - |L : Z| - \sum_{j \geq i} \chi_j(1)^2.$$

For all j, $\theta_j(1)$ is a power of p, and so $\theta_i(1)^2$ divides $\sum_{j \geq i} \chi_j(1)^2$. By [Is], Corollary 2.30, $\theta_i(1)^2 \leq |K : Z|$, and so $\theta_i(1)^2$ divides $|L| - |L : Z|$. It follows that $\theta_i(1)^2$ divides $\sum_{1 \leq j < i} \chi_j(1)^2$. By (6.4.c), $|L : K|$ is prime to p, and so $\chi_i(1)^2$ divides $\sum_{1 \leq j < i} \chi_j(1)^2$. Then

$$2\chi_i(1)\chi_1(1) < p\chi_i(1)\chi_1(1) \leq \chi_i(1)^2 \leq \sum_{j < i} \chi_j(1)^2.$$

Repeated use of Theorem (5.6) then shows that $\mathcal{X}$ is coherent. $\square$

(6.7) *Let G be a finite group, p a prime number, P a Sylow p-subgroup of G and $L = N_G(P)$. Assume that $|L|$ is odd and that $P^{\#}$ is a TI-subset of G. Let Z be a normal subgroup of L such that $1 \neq Z \subset Z(P)$ and such that $|C_L(z)|$ is independent of z for $z \in Z^{\#}$. Let $\psi \in \operatorname{Irr} G$ be such that ψ is constant on $Z^{\#}$. Then, for $z \in Z^{\#}$, $\psi(z) \in \mathbf{Z}$ and $\psi(z) \equiv \psi(1) \pmod{|P|}$.*

Proof. Since $(\operatorname{Res}_Z^G \psi, 1_Z) \in \mathbf{N}$, $\psi(z) \in \mathbf{Q}$, and, since $\psi(z)$ is an algebraic integer, $\psi(z) \in \mathbf{Z}$. The notation

$$\alpha \equiv \beta \pmod{|P|}$$

means that α, β and $(\alpha - \beta)/|P|$ are algebraic integers. Let C_s, $0 \leq s \leq n$, be the conjugacy classes of G, and let C_s be the sum of the elements of C_s taken in $\mathbf{C}[G]$. Let $\omega : Z\mathbf{C}[G] \to \mathbf{C}$ be the $\mathbf{C}$-linear mapping such that $\omega(C_s) = \psi(C_s)/\psi(1)$ for all s. We know that ω is an algebra homomorphism ([Is], p. 35). For $0 \leq i, j, s \leq n$, let $a_{ijs} \in \mathbf{N}$ be such that $C_i C_j = \sum_{s=0}^n a_{ijs} C_s$.

(6.7.1) *Let i, j and s be such that $C_i \cap Z^{\#} \neq \emptyset$, $C_j \cap Z^{\#} \neq \emptyset$ and $C_s \cap Z = \emptyset$. Then $a_{ijs}|C_s| \equiv 0 \pmod{|P|}$.*

Proof. We show that P acts fixed-point-freely on the set

$$\Omega = \{(u, v) \in C_i \times C_j \mid uv \in C_s\}$$

of cardinality $a_{ijs}|C_s|$. Let $x \in P^{\#}$ and $(u, v) \in \Omega$ be such that $(u, v)^x = (u, v)$. Let $y \in \{u, v\}$. Since $P^{\#}$ is a TI-subset, $C_G(x) \subset L$. Thus y is a p-element of

L, and so $y \in P$. By hypothesis there is an element $g \in G$ for which $y^g \in Z^\#$. Since $P^\#$ is a TI-subset, $g \in L$ and, since Z is normal in L, $y \in Z$. It follows that $uv \in Z$, which contradicts the hypothesis. $\qquad\square$

Set $\alpha = \omega(C_s)$ for some index s for which $C_s \cap Z^\# \neq \emptyset$. Since $\psi(z)$ and $|C_L(z)|$ are independent of z for $z \in Z^\#$, α does not depend on s. Let $C_0 = \{1\}$.

(6.7.2) *Let i and j be indices such that $C_i \cap Z^\# \neq \emptyset$ and $C_j \cap Z^\# \neq \emptyset$. Then*

$$\psi(1)\alpha^2 \equiv \psi(1)(a_{ij0} + a_{ij}\alpha) \pmod{|P|},$$

where $a_{ij} = \sum_{s, C_s \cap Z^\# \neq \emptyset} a_{ijs}$.

Proof. Since $\omega(C_i)\omega(C_j) = \sum_{s=0}^{n} a_{ijs}\omega(C_s)$, it follows that

$$\psi(1)\alpha^2 = \sum_{s=0}^{n} \psi(1)a_{ijs}\omega(C_s).$$

If $C_s \cap Z = \emptyset$ and $x \in C_s$, then $\psi(1)a_{ijs}\omega(C_s) = a_{ijs}|C_s|\psi(x) \equiv 0 \pmod{|P|}$ by (6.7.1). Thus,

$$\psi(1)\alpha^2 \equiv \sum_{s, C_s \cap Z \neq \emptyset} \psi(1)a_{ijs}\omega(C_s) = \psi(1)(a_{ij0} + a_{ij}\alpha) \pmod{|P|}.$$

$\qquad\square$

(6.7.3) *If $z \in Z^\#$, then $\psi(z) \equiv \psi(1) \pmod{|P|}$.*

Proof. Since $|L|$ is odd, z^{-1} is not conjugate to z in G. We may then assume that $z \in C_1$ and $z^{-1} \in C_2$. Then $a_{110} = 0$ and $a_{120} = |C_1|$. By (6.7.2) applied to $(i, j) = (1, 1)$ and then to $(i, j) = (1, 2)$,
$$\psi(1)a_{11}\alpha \equiv \psi(1)(|C_1| + a_{12}\alpha) \pmod{|P|}.$$
Since $\psi(1)\alpha = |C_1|\psi(z)$ and $|C_1|$ is prime to p, it follows that
$$a_{11}\psi(z) \equiv \psi(1) + a_{12}\psi(z) \pmod{|P|}.$$
In particular, $a_{11} \equiv 1 + a_{12} \pmod{|P|}$ for $\psi = 1_G$, and so
$$(1 + a_{12})\psi(z) \equiv \psi(1) + a_{12}\psi(z) \pmod{|P|},$$
whence $\psi(z) \equiv \psi(1) \pmod{|P|}$. $\qquad\square$

(6.8) Theorem. *Let G be a finite group and let L be a subgroup of G. Assume that:*

(a) *$L = H \rtimes W_1$, $|L|$ is odd, H is a non-identity nilpotent subgroup of L and $H^\#$ is a TI-subset of G with normalizer L.*

(b) *$S = \{\mathrm{Ind}_H^L \theta \mid \theta \in \mathrm{Irr}\, H,\ \theta \neq 1_H\}$ and τ is the restriction to $\mathbf{Z}[S, L^\#]$ of Ind_L^G.*

(c) *One of the following two cases holds.*

    (c1) *L is a Frobenius group with kernel H.*

    (c2) *Hypothesis (4.6) holds with $H = K$ and $A = H^\#$, w_2 is prime and $W_2 \subset [H, H]$.*

Then S is coherent.

Proof. Since $|L|$ is odd, (1.5.e) shows that (5.2.a) is satisfied. By the definition of $\mathcal{S}$, $\mathbf{Z}[\mathcal{S}, L^{\#}] = \mathbf{Z}[\mathcal{S}, H^{\#}]$. Since $H^{\#}$ is a TI-subset with normalizer L, (5.2.b) holds by [Is], Lemma 7.7. Moreover, (2.3), Definition (2.5) and [Is], Lemma 7.7, show that τ coincides with the Dade isometry relative to (A, L, G) in case (c2). By (5.3) and [Is], Theorem 6.34, we see that Hypothesis (5.2) holds. Consequently, Hypothesis (6.4) holds with $K = H$ and $M = 1$. By (6.5), we may assume that H is a non-abelian p-group for some prime number p. Denote $[H, H]$ by H'; in case (c1), set $W_2 = 1$; $\mathcal{S}(X)$ has the same meaning as in Hypothesis (6.1). We consider two cases.

(A) $Z(H) \cap W_2 = 1$.

(B) $1 \neq W_2 \subset Z(H)$.

Set $Z = Z(H) \cap H'$ in case (A) and $Z = W_2$ in case (B). Let $\mathcal{X} = \mathcal{S} - \mathcal{S}(Z)$ and $\mathcal{Y} = \mathcal{S}(H')$. As $Z \subset H'$, $\mathcal{X} \cap \mathcal{Y} = \emptyset$. Set $\mathcal{Y} = \{\eta_1, \ldots, \eta_m\}$, where $|\mathcal{Y}| = m$. By (c), (1.6) and [Is], Theorem 6.34, $\mathcal{Y} \subset \operatorname{Irr} L$ and $\eta_j(1) = |W_1|$ for all j. By (1.1) and (1.4), $\mathcal{Y}$ is coherent. Let τ_1 be an isometry from $\mathbf{Z}[\mathcal{Y}]$ to $\mathbf{Z}[\operatorname{Irr} G]$ which coincides with τ on $\mathbf{Z}[\mathcal{Y}, L^{\#}]$.

(6.8.1) $\mathcal{X} \cup \mathcal{Y}$ *is coherent in case* (A).

Proof. By [Is], Theorem 6.34, $\mathcal{X} \subset \operatorname{Irr} L$ in case (c1). Suppose that (c2) holds. Then L and L/Z satisfy Hypothesis (4.2) and, for $x \in W_1^{\#}$, $|C_H(x)| = |C_{H/Z}(x)|$ because $Z \cap W_2 = 1$. By (1.6) and Theorem (4.5), $\mathcal{S}$ and $\mathcal{S}(Z)$ each have $w_2 - 1$ reducible characters. Thus, $\mathcal{X} \subset \operatorname{Irr} L$. In both cases,

$$\mathcal{X} = \{\chi \in \operatorname{Irr} L \mid Z \not\subseteq \operatorname{Ker} \chi\}$$

and $\mathcal{X}$ is coherent by (6.6).

Let $\mathcal{X} = \{\chi_1, \ldots, \chi_n\}$, where $|\mathcal{X}| = n$, $\chi_1(1) \leq \cdots \leq \chi_n(1)$ and $\chi_i(1) = d_i \chi_1(1)$ for $1 \leq i \leq n$. Since H is a p-group, $d_i \in \mathbf{N}$ for all i. Let τ_2 be an isometry from $\mathbf{Z}[\mathcal{X}]$ to $\mathbf{Z}[\operatorname{Irr} G]$ which coincides with τ on $\mathbf{Z}[\mathcal{X}, L^{\#}]$. Since $(\eta_j - \eta_1, \chi_i - d_i\chi_1) = 0$ for all i and j, $\mathcal{Y}^{\tau_1}$ is orthogonal to $\mathcal{X}^{\tau_2}$ by (4.1). Set $\chi_1(1) = a|W_1|$. Since $((\chi_1 - a\eta_1)^\tau, (\eta_j - \eta_1)^\tau) = a$ for $j > 1$, we can write

$$(\chi_1 - a\eta_1)^\tau = X - a\eta_1^{\tau_1} + b\sum_{j=1}^{m} \eta_j^{\tau_1},$$

where $b \in \mathbf{Z}$ and X is orthogonal to $\mathcal{Y}^{\tau_1}$. Since $(\operatorname{Res}_L^G(\eta_1^{\tau_1}), \chi_i - d_i\chi_1) = (\eta_1^{\tau_1}, \chi_i^{\tau_2} - d_i\chi_1^{\tau_2}) = 0$ for all i, we can write

$$\operatorname{Res}_L^G(\eta_1^{\tau_1}) = c\sum_{i=1}^{n} d_i\chi_i + \chi',$$

where $c \in \mathbf{Z}$ and where χ' is orthogonal to $\mathcal{X}$ and so constant on Z. Then

$$(\operatorname{Res}_L^G(\eta_1^{\tau_1}), \chi_1 - a\eta_1) = \left(\eta_1^{\tau_1}, X - a\eta_1^{\tau_1} + b\sum_j \eta_j^{\tau_1}\right) = b - a,$$

and so $c = (\mathrm{Res}_L^G(\eta_1^{\tau_1}), \chi_1) \equiv b \pmod{a}$. Since $\mathcal{X} = \{\chi \in \mathrm{Irr}\, L \mid Z \not\subseteq \mathrm{Ker}\, \chi\}$, we see that

$$\sum_{i=1}^n d_i \chi_i = \frac{1}{a|W_1|} \sum_{i=1}^n \chi_i(1)\chi_i = \frac{1}{a|W_1|}(\rho_L - \rho_{L/Z}),$$

where ρ_M denotes the regular character of the group M. If $z \in Z^\#$, it follows that

$$\eta_1^{\tau_1}(z) - \eta_1^{\tau_1}(1) = c\Big(\sum_i d_i \chi_i(z) - \sum_i d_i \chi_i(1)\Big) = \frac{-c|H|}{a}.$$

Thus $\eta_1^{\tau_1}$ is constant on $Z^\#$. By (6.7), $\dfrac{-c|H|}{a} \equiv 0 \pmod{|H|}$; therefore, $b \equiv c \equiv 0 \pmod{a}$. Set $b = ax$. Then

$$1 + a^2 = \|\chi_1 - a\eta_1\|^2 = \|X\|^2 + a^2(x-1)^2 + (m-1)x^2 a^2,$$

and so $(x-1)^2 + (m-1)x^2 \leq 1 + 1/a^2$. Since $\mathcal{X} \cap \mathcal{Y} = \emptyset$, $a > 1$; thus, $x = 0$, or else $x = 1$ and $m = 2$. The second case reduces to the first on replacing $\eta_1^{\tau_1}$ and $\eta_2^{\tau_1}$ by $-\eta_2^{\tau_1}$ and $-\eta_1^{\tau_1}$ respectively. Thus we may assume that $b = 0$. Then $(\chi_1 - a\eta_1)^\tau = X - a\eta_1^{\tau_1}$ and $\|X\|^2 = \|\chi_1\|^2 = 1$. Considering $((\chi_1 - a\eta_1)^\tau, (\chi_2 - \chi_1)^\tau)$, we see that $X = \chi_1^{\tau_2}$ or $X = -\chi_2^{\tau_2}$. If $n \geq 3$, we see that $X = \chi_1^{\tau_2}$ by considering $((\chi_1 - a\eta_1)^\tau, (\chi_3 - d_3\chi_1)^\tau)$. If $n = 2$, we may assume that $X = \chi_1^{\tau_2}$, possibly on replacing $\chi_1^{\tau_2}$ and $\chi_2^{\tau_2}$ by $-\chi_2^{\tau_2}$ and $-\chi_1^{\tau_2}$ respectively. If τ_3 is the $\mathbf{Z}$-linear mapping from $\mathbf{Z}[\mathcal{X} \cup \mathcal{Y}]$ to $\mathbf{Z}[\mathrm{Irr}\, G]$ which coincides with τ_1 on $\mathcal{Y}$ and with τ_2 on $\mathcal{X}$, then τ_3 is an isometry which coincides with τ on $\mathbf{Z}[\mathcal{X} \cup \mathcal{Y}, L^\#]$. $\qquad\square$

(6.8.2) $\mathcal{X} \cup \mathcal{Y}$ *is coherent in case* (B).

Suppose that case (B) holds.

(6.8.2.1) *If $\eta \in \mathcal{Y}$, then η^{τ_1} is constant on $Z^\#$.*

Proof. Let $x \in Z^\#$ and let k be such that $1 \leq k < w_2$. Since w_2 is prime, by (1.9) there is an automorphism u of $\mathbf{Q}_{|G|}$ such that $\eta^{\tau_1 u}(x) = \eta^{\tau_1}(x^k)$. By (5.9.a), $\eta^{\tau_1 u} = \eta^{u\tau_1}$. Since $Z \subset H' \subset \mathrm{Ker}\, \eta$, $\eta^u - \eta$ vanishes on Z. By definition of τ, $\eta^{u\tau_1}(x) - \eta^{\tau_1}(x) = (\eta^u - \eta)^\tau(x) = 0$, and so $\eta^{\tau_1}(x^k) = \eta^{\tau_1}(x)$. $\qquad\square$

(6.8.2.2) *Let $\varphi \in \mathrm{Irr}\, Z$, $\varphi \neq 1_Z$. Then*

$$(\mathrm{Ind}_Z^L \varphi - |H : Z|\eta_1)^\tau = X - |H : Z|Y,$$

where X is orthogonal to $\mathcal{Y}^{\tau_1}$ and $Y = \eta_1^{\tau_1}$, or else $m = 2$ and $Y = -\eta_2^{\tau_1}$. Moreover, Y is independent of φ.

Proof. Let $\alpha = \mathrm{Ind}_Z^L \varphi - |H : Z|\eta_1$. Thus, $\mathrm{Supp}(\alpha) \subset H^\#$. If $\psi \in \mathcal{Y}^{\tau_1}$, then

$$(\alpha^\tau, \psi) = (\alpha, \mathrm{Res}_L^G \psi) = (\varphi, \mathrm{Res}_Z^G \psi) - |H : Z|(\eta_1, \mathrm{Res}_L^G \psi).$$

By (6.8.2.1), there are integers $a, b \in \mathbf{Z}$ such that $\operatorname{Res}_Z^G \psi = a\rho_Z + b1_Z$, where ρ_Z is the regular character of Z, and $a = (\varphi, \operatorname{Res}_Z^G \psi)$. If $z \in Z^\#$, by (6.7), $b = \psi(z) \equiv \psi(1) = a|Z| + b \pmod{|H|}$. Thus,

$$(\alpha^\tau, \psi) = a - |H : Z|(\eta_1, \operatorname{Res}_L^G \psi) \equiv 0 \pmod{|H : Z|},$$

and (α^τ, ψ) is independent of φ. For $j > 1$, $(\alpha^\tau, \eta_j^{\tau_1} - \eta_1^{\tau_1}) = (\alpha, \eta_j - \eta_1) = |H : Z|$. It follows that

$$\alpha^\tau = X - |H : Z|\eta_1^{\tau_1} + x|H : Z|\sum_{j=1}^m \eta_j^{\tau_1},$$

where $x \in \mathbf{Z}$ and X is orthogonal to $\mathcal{Y}^{\tau_1}$. By (1.5.b),

$$\|\alpha^\tau\|^2 = \|\alpha\|^2 = |I_L(\varphi) : Z| + |H : Z|^2 = |L : Z| + |H : Z|^2.$$

Also, $|L : Z| = |W_1||H : Z| < |H : Z|^2$ since W_1 acts fixed-point-freely on $H/Z \neq 1$. Thus, $\|\alpha^\tau\|^2 < 2|H : Z|^2$. It follows that $(x - 1)^2 + (m - 1)x^2 \leq 1$. Thus, $x = 0$, or else $x = 1$ and $m = 2$. $\qquad\square$

(6.8.2.3) *Let* $\chi \in \mathcal{X}$ *and* $a = \chi(1)/|W_1|$. *Then* $(\chi - a\eta_1)^\tau = X_1 - aY$, *where* X_1 *is orthogonal to* $\mathcal{Y}^{\tau_1}$ *and* Y *is the same as in* (6.8.2.2).

Proof. Let $\chi = \operatorname{Ind}_H^L \theta$ where $\theta \in \operatorname{Irr} H$ with $Z \not\subset \operatorname{Ker} \theta$. By [Is], Lemma 2.27, since $Z \subset Z(H)$, there is a character $\varphi \in \operatorname{Irr} Z$ such that $\varphi \neq 1_Z$ and $\operatorname{Res}_Z^H \theta = a\varphi$. Set $\operatorname{Ind}_Z^H \varphi = \sum_{i=1}^r a_i\theta_i$ with $\theta_i \in \operatorname{Irr} H$ pairwise distinct and $a_i \neq 0$. Then $\operatorname{Res}_Z^H \theta_i = a_i\varphi$ with $\theta_i(1) = a_i$; we may assume that $\theta = \theta_1$ and that $a = a_1$. Set $\chi_i = \operatorname{Ind}_H^L \theta_i$ and $\alpha_i = \chi_i - a_i\eta_1$ $(1 \leq i \leq r)$. It follows that $\operatorname{Supp}(\alpha_i) \subset H^\#$; also,

$$\operatorname{Ind}_Z^L \varphi - |H : Z|\eta_1 = \sum_{i=1}^r a_i\alpha_i \quad \text{and} \quad \sum_{i=1}^r a_i^2 = |H : Z|.$$

Indeed, $\operatorname{Ind}_Z^L \varphi - |H : Z|\eta_1 - \sum_{i=1}^r a_i\alpha_i = -|H : Z|\eta_1 + \sum_{i=1}^r a_i^2\eta_1$ and this difference vanishes on 1. By (5.3) and (5.5), $R(\chi_i)$ is orthogonal to $\mathcal{Y}^{\tau_1}$ in the notation of Hypothesis (5.2). Set $\alpha_i^\tau = X_i - b_iY + Z_i$, where $X_i \in \mathbf{Z}[R(\chi_i)]$ and Z_i is orthogonal to $R(\chi_i)$ and to Y. By (5.4.a), $\|X_i\|^2 \geq \|\chi_i\|^2$. Thus,

$$b_i^2 \leq \|b_iY - Z_i\|^2 = \|\alpha_i\|^2 - \|X_i\|^2 = \|\chi_i\|^2 + a_i^2 - \|X_i\|^2 \leq a_i^2.$$

As $a_i > 0$, $b_i \leq a_i$. By (6.8.2.2), $\sum_{i=1}^r a_i\alpha_i^\tau = X - |H : Z|Y$, and so

$$|H : Z| = \sum_{i=1}^r a_ib_i \leq \sum_{i=1}^r a_i^2 = |H : Z|.$$

It follows that $b_i = a_i$ for all i. Then $\| - b_iY + Z_i\|^2 \geq a_i^2$ and, by (5.4.b), $\alpha_i^\tau = X_i - a_iY$ for all i. $\qquad\square$

Proof of (6.8.2). Let τ_2 be the $\mathbf{Z}$-linear mapping from $\mathbf{Z}[\mathcal{X} \cup \mathcal{Y}]$ to $\mathbf{Z}[\operatorname{Irr} G]$ which coincides with τ on $\mathbf{Z}[\mathcal{X} \cup \mathcal{Y}, L^\#]$ and which satisfies $\eta_1^{\tau_2} = Y$. By (6.8.2.3), $((\chi - a\eta_1)^\tau, \eta_1^{\tau_2}) = (\chi - a\eta_1, \eta_1)$ if $\chi \in \mathcal{X}$ and $a = \dfrac{\chi(1)}{\eta_1(1)}$. We see that $((\eta - \eta_1)^\tau, \eta_1^{\tau_2}) = (\eta - \eta_1, \eta_1)$ if $\eta \in \mathcal{Y}$. Thus τ_2 preserves the inner product of any two elements of $\mathbf{Z}[\mathcal{X} \cup \mathcal{Y}, L^\#] \cup \{\eta_1\}$, which generates $\mathbf{Z}[\mathcal{X} \cup \mathcal{Y}]$. $\square$

(6.8.3) $\mathcal{S}$ *is coherent.*

Proof. Suppose that $\mathcal{S}$ is not coherent. By (6.8.1) and (6.8.2), $\mathcal{S} \neq \mathcal{X} \cup \mathcal{Y}$ and so $Z \neq H'$. There is a set $\mathcal{S}_1$, closed under complex conjugation, such that $\mathcal{X} \cup \mathcal{Y} \subset \mathcal{S}_1 \subset \mathcal{S}$, and there is a set $\mathcal{S}_2 = \{\psi, \overline{\psi}\} \subset \mathcal{S}$ such that $\mathcal{S}_1$ is coherent but $\mathcal{S}_1 \cup \mathcal{S}_2$ is not. By Theorem (5.6), it follows that

$$2\psi(1)\eta_1(1) \geq \sum_{\chi \in \mathcal{S}_1} \frac{\chi(1)^2}{\|\chi\|^2} > \sum_{\chi \in \mathcal{X}} \frac{\chi(1)^2}{\|\chi\|^2}.$$

By (1.5.c, d),

$$
\begin{aligned}
\sum_{\chi \in \mathcal{X}} \frac{\chi(1)^2}{\|\chi\|^2} \;&=\; \sum_{\substack{\theta \,\in\, \operatorname{Irr} H \\ Z \,\not\subset\, \operatorname{Ker} \theta}} |L : H|\theta(1)^2 \\
&=\; |L : H|(|H| - |H : Z|) \\
&=\; |W_1||H : Z|(|Z| - 1).
\end{aligned}
$$

If ψ is induced to L from an irreducible character of H of degree d, then

$$2d|W_1|^2 > |W_1||H : Z|(|Z| - 1).$$

By [Is], Corollary 2.30, $d^2 \leq |H : Z|$, and so

$$4|W_1|^2 > |H : Z|(|Z| - 1)^2.$$

But, in case (A), W_1 acts fixed-point-freely on Z and $|Z|$ is odd so that $|Z| - 1 \geq 2|W_1|$; on the other hand, in case (B), W_1 acts fixed-point-freely on H/H' and on H'/Z so that $|H : Z| \geq (2|W_1| + 1)^2$. In each case, we obtain a contradiction. $\square$

7. Non-existence of a Certain Type of Group of Odd Order

(7.1) Hypothesis. *Assume Hypothesis (2.2) and denote by τ the Dade isometry relative to (A, L, G). For $\chi \in \mathrm{CF}(G)$, denote by χ^ρ the element of $\mathrm{CF}(L, A)$ for which*

$$\chi^\rho(a) = \frac{1}{|H(a)|} \sum_{x \in H(a)} \chi(ax)$$

for $a \in A$. Set $A^\tau = \bigcup_{a \in A} (aH(a))^G$.

(7.2) *Assume Hypothesis (7.1).*

(a) *If $\alpha \in \mathrm{CF}(L, A)$, then $\alpha^{\tau\rho} = \alpha$.*

(b) *If $\chi \in \mathrm{CF}(G)$, then $\|\chi^\rho\|^2 \leq \|\chi\|^2$, with equality if and only if χ is in the image of τ.*

Proof. (a) For $a \in A$,

$$\alpha^{\tau\rho}(a) = \frac{1}{|H(a)|} \sum_{x \in H(a)} \alpha^\tau(ax) = \frac{1}{|H(a)|} \sum_{x \in H(a)} \alpha(a) = \alpha(a).$$

(b) By (a), $\chi^{\rho\tau\rho} = \chi^\rho$. Then, by (2.7), $(\alpha^\tau, \chi^{\rho\tau} - \chi) = (\alpha, \chi^{\rho\tau\rho} - \chi^\rho) = 0$ for all $\alpha \in \mathrm{CF}(L, A)$. Thus $\chi^{\rho\tau}$ is the orthogonal projection of χ on the image of τ. It follows that $\|\chi\|^2 \geq \|\chi^{\rho\tau}\|^2$, with equality if and only if χ is in the image of τ. Moreover, $\|\chi^{\rho\tau}\|^2 = \|\chi^\rho\|^2$ by Theorem (2.6). $\square$

(7.3) *Assume Hypothesis (7.1). Let $\chi \in \mathrm{CF}(G)$. Then*

$$\frac{1}{|G|} \sum_{g \in A^\tau} |\chi(g)|^2 \geq \frac{1}{|L|} \sum_{a \in A} |\chi^\rho(a)|^2 = \|\chi^\rho\|^2.$$

Equality holds if and only if χ is constant on $aH(a)$ for all $a \in A$.

Proof. Let $\chi_1 \in \mathrm{CF}(G)$ be defined by setting

$$\chi_1(g) = \begin{cases} \chi(g) & \text{if } g \in A^\tau, \\ 0 & \text{if } g \notin A^\tau. \end{cases}$$

By the definition of ρ, $\chi_1^\rho = \chi^\rho$. By (7.2.b),

$$\frac{1}{|G|} \sum_{g \in A^\tau} |\chi(g)|^2 = \|\chi_1\|^2 \geq \|\chi_1^\rho\|^2 = \|\chi^\rho\|^2.$$

Equality holds if and only if χ_1 is in the image of τ, that is, if and only if, for all $a \in A$, χ is constant on $aH(a)$. $\square$

(7.4) Hypothesis. *Let G be a finite group and let $(L_i)_{i \in I}$ be a family of subgroups of G. Assume that, for all $i \in I$, Hypothesis (7.1) is satisfied with $(A_i, L_i, \tau_i, \rho_i)$ in place of (A, L, τ, ρ). Assume that the sets $A_i^{\tau_i}$ are pairwise disjoint, and set $G_0 = G - \bigcup_{i \in I} A_i^{\tau_i}$.*

(7.5) *Assume Hypothesis (7.4). Let $\chi \in \operatorname{Irr} G$. Then*

$$\frac{1}{|G|} \left(\sum_{g \in G_0} |\chi(g)|^2 - |G_0| \right) + \sum_{i \in I} \left(\|\chi^{\rho_i}\|^2 - \frac{|A_i|}{|L_i|} \right) \leq 0.$$

Proof. By (7.3),

$$1 = \|\chi\|^2 \;=\; \frac{1}{|G|} \sum_{g \in G_0} |\chi(g)|^2 + \sum_{i \in I} \frac{1}{|G|} \sum_{g \in A_i^{\tau_i}} |\chi(g)|^2$$

$$\geq \; \frac{1}{|G|} \sum_{g \in G_0} |\chi(g)|^2 + \sum_{i \in I} \|\chi^{\rho_i}\|^2.$$

Also, by (7.3) applied to $\chi = 1_G$,

$$1 = \frac{|G_0|}{|G|} + \sum_{i \in I} \|1_G^{\rho_i}\|^2 = \frac{|G_0|}{|G|} + \sum_{i \in I} \frac{|A_i|}{|L_i|}.$$

Thus, $\displaystyle \frac{1}{|G|} \sum_{g \in G_0} |\chi(g)|^2 + \sum_{i \in I} \|\chi^{\rho_i}\|^2 \leq \frac{|G_0|}{|G|} + \sum_{i \in I} \frac{|A_i|}{|L_i|}.$ $\qquad\square$

(7.6) Hypothesis. *Assume Hypothesis (7.1). Assume that there is a normal subgroup H of L such that $A = H^{\#}$. Set $|H| = h$ and $|L : H| = e$. Let $T = \{\operatorname{Ind}_H^L \theta \mid \theta \in \operatorname{Irr} H\} = \{\zeta_0, \ldots, \zeta_n\}$, where the characters ζ_i are pairwise distinct, and let $\zeta_i(1) = d_i \zeta_0(1)$ $(1 \leq i \leq n)$.*

(7.7) *Assume Hypothesis (7.6). Let $\chi \in \operatorname{CF}(G)$ and set $c_i = ((\zeta_i - d_i \zeta_0)^\tau, \chi)$ for $1 \leq i \leq n$.*

(a) *For $x \in A$, $\displaystyle \chi^\rho(x) = \sum_{i=1}^{n} \frac{\overline{c_i}}{\|\zeta_i\|^2} \zeta_i(x).$*

(b) $\displaystyle \|\chi^\rho\|^2 = \sum_{i=1}^{n} \sum_{j=1}^{n} \frac{\overline{c_i} c_j}{\|\zeta_i\|^2 \|\zeta_j\|^2} \left((\zeta_i, \zeta_j) - \frac{\zeta_i(1) \zeta_j(1)}{eh} \right).$

Proof. (a) Let $\psi_i = \zeta_i - d_i \zeta_0$ for $1 \leq i \leq n$. Then $\psi_i \in \operatorname{CF}(L, A)$ and, by (2.7), $c_i = (\psi_i, \chi^\rho)$. Let $\psi \in \operatorname{CF}(L, A)$. For $x \in H$,

$$(\operatorname{Ind}_H^L \operatorname{Res}_H^L \psi)(x) = \frac{1}{h} \sum_{g \in L} \psi(gxg^{-1}) = e\psi(x).$$

Thus, $\psi = (1/e)\operatorname{Ind}_H^L \operatorname{Res}_H^L \psi$ is a **C**-linear combination of elements of T. As $\psi(1) = 0$, we see that ψ is a **C**-linear combination of the functions ψ_i,

$1 \leq i \leq n$. It follows that, for $b_i \in \mathbb{C}$, $\chi^\rho - \sum_{i=0}^n b_i \zeta_i$ vanishes on A if $(\psi_j, \chi^\rho - \sum_{i=0}^n b_i \zeta_i) = 0$ for all j. This condition holds if

$$c_j = (\psi_j, \chi^\rho) = (\zeta_j - d_j \zeta_0, \sum b_i \zeta_i) = \overline{b_j} \|\zeta_j\|^2 - d_j \overline{b_0} \|\zeta_0\|^2$$

for all j. This occurs if $b_0 = 0$ and $b_j = \overline{c_j}/\|\zeta_j\|^2$ for $j \geq 1$.

(b) By (a),

$$\|\chi^\rho\|^2 = \frac{1}{eh} \sum_{x \in A} \left| \sum_{i=1}^n \frac{\overline{c_i}}{\|\zeta_i\|^2} \zeta_i(x) \right|^2$$

$$= \frac{1}{eh} \sum_{i,j \geq 1} \frac{\overline{c_i} c_j}{\|\zeta_i\|^2 \|\zeta_j\|^2} \sum_{x \in A} \zeta_i(x) \overline{\zeta_j(x)}.$$

But

$$(\zeta_i, \zeta_j) = \frac{1}{eh} \sum_{x \in L} \zeta_i(x) \overline{\zeta_j(x)} = \frac{\zeta_i(1)\zeta_j(1)}{eh} + \frac{1}{eh} \sum_{x \in A} \zeta_i(x) \overline{\zeta_j(x)}.$$

Thus,

$$\frac{1}{eh} \sum_{x \in A} \zeta_i(x) \overline{\zeta_j(x)} = (\zeta_i, \zeta_j) - \frac{\zeta_i(1)\zeta_j(1)}{eh}.$$

$\square$

(7.8) *Assume Hypothesis (7.6). Suppose that $S = T - \{\mathrm{Ind}_H^L 1_H\}$ is coherent. Let ν be an extension of τ to an isometry from $\mathbf{Z}[S]$ to $\mathbf{Z}[\mathrm{Irr}\, G]$. Also, let $\zeta \in S \cap \mathrm{Irr}\, L$ be such that $\zeta(1) = e$. Let $\beta = (\mathrm{Ind}_H^L 1_H - \zeta)^\tau$.*

(a) *S^ν is orthogonal to 1_G. There is an integer a and a function $\Gamma \in \mathrm{CF}(G)$ orthogonal to $S^\nu \cup \{1_G\}$ for which*

$$\beta = 1_G - \zeta^\nu + a \sum_{\varphi \in S} \frac{\varphi(1)}{e\|\varphi\|^2} \varphi^\nu + \Gamma.$$

(b) *Suppose that $e \leq \dfrac{h-1}{2}$. Then $\|\zeta^{\nu\rho}\|^2 \geq 1 - \dfrac{e}{h}$ and $\|\Gamma\|^2 \leq e - 1$.*

(c) *Let $\chi \in \mathrm{Irr}\, G$ be such that χ is orthogonal to S^ν. Then $\chi^\rho(x) = (\beta, \chi)$ for $x \in A$ and $\|\chi^\rho\|^2 = \dfrac{|A|}{|L|}(\beta, \chi)^2$.*

Proof. (a) Since S is coherent, $|S| \geq 2$ by Definition (5.1). Let $\varphi \in S$, $\varphi \neq \zeta$. By (2.7),

$$\left(\varphi^\nu - \frac{\varphi(1)}{e}\zeta^\nu, 1_G\right) = \left((\varphi - \frac{\varphi(1)}{e}\zeta)^\tau, 1_G\right) = \left(\varphi - \frac{\varphi(1)}{e}\zeta, 1_L\right) = 0.$$

As $\|\zeta^\nu\| = 1$ and $(\varphi^\nu, \zeta^\nu) = 0$ by (1.5.c), it follows that

$$(\varphi^\nu, 1_G) = \frac{\varphi(1)}{e}(\zeta^\nu, 1_G) = 0.$$

If $\theta \in \operatorname{Irr} H$ is such that $\zeta = \operatorname{Ind}_H^L \theta$, then $(\beta, 1_G) = (\operatorname{Ind}_H^L 1_H - \zeta, 1_L) = (1_H - \theta, 1_H) = 1$ by (2.7). We can write $\beta = 1_G - \zeta^\nu + \sum_{\varphi \in \mathcal{S}} a_\varphi \varphi^\nu + \Gamma$ where $a_\varphi \in \mathbf{C}$ and $\Gamma \in \operatorname{CF}(G)$ is orthogonal to $\mathcal{S} \cup \{1_G\}$. By (1.5.c), the elements of $\mathcal{S}$ are pairwise orthogonal. For $\varphi \in \mathcal{S} - \{\zeta\}$,

$$\left(\beta, \varphi^\nu - \frac{\varphi(1)}{e}\zeta^\nu\right) = \left(\operatorname{Ind}_H^L 1_H - \zeta, \varphi - \frac{\varphi(1)}{e}\zeta\right) = \frac{\varphi(1)}{e}.$$

Thus,

$$a_\varphi \|\varphi\|^2 - \frac{\varphi(1)}{e}(a_\zeta - 1) = \frac{\varphi(1)}{e}$$

and we have the required expression, with $a = a_\zeta$. Moreover, $(\beta, \zeta^\nu) = -1 + a$ and so $a \in \mathbf{Z}$.

(b) We use (7.7) with $\zeta_0 \in \mathcal{S} - \{\zeta\}$, $\zeta_1 = \operatorname{Ind}_H^L 1_H$, $\zeta_2 = \zeta$ and $\chi = \zeta^\nu$. Thus,

$$c_1 = (\beta + (\zeta - d_1 \zeta_0)^\tau, \zeta^\nu) = a - 1 + 1 = a,$$
$$c_2 = ((\zeta - d_2 \zeta_0)^\tau, \zeta^\nu) = 1$$

and, for $i > 2$,

$$c_i = ((\zeta_i - d_i \zeta_0)^\tau, \zeta^\nu) = 0.$$

By (7.7.b), we have $\|\zeta^{\nu\rho}\|^2 = ua^2 - 2va + w$, where

$$u = \frac{1}{\|\zeta_1\|^4}\left(\|\zeta_1\|^2 - \frac{\zeta_1(1)^2}{eh}\right) = \frac{1}{e}\left(1 - \frac{1}{h}\right),$$

$$v = \frac{1}{\|\zeta_1\|^2\|\zeta_2\|^2}\frac{\zeta_1(1)\zeta_2(1)}{eh} = \frac{1}{h}$$

and

$$w = \frac{1}{\|\zeta_2\|^4}\left(\|\zeta_2\|^2 - \frac{\zeta_2(1)^2}{eh}\right) = 1 - \frac{e}{h}.$$

On the other hand,

$$e + 1 = \|\beta\|^2 = 1 + (a-1)^2 + a^2 \sum_{\varphi \in \mathcal{S} - \{\zeta\}} \frac{\varphi(1)^2}{e^2\|\varphi\|^2} + \|\Gamma\|^2.$$

By (1.5.d), $\displaystyle\sum_{\varphi \in \mathcal{S}} \frac{\varphi(1)^2}{e^2\|\varphi\|^2} = \frac{e}{e^2} \sum_{\theta \in \operatorname{Irr} H - \{1_H\}} \theta(1)^2 = \frac{1}{e}(h - 1) = hu.$ Thus,

$$\|\Gamma\|^2 = e - (a-1)^2 - a^2(hu - 1) = e - 1 - h(ua^2 - 2va).$$

By hypothesis, $2hv = 2 \leq (h-1)/e = hu$, and so $0 \leq 2v/u \leq 1$. As $a \in \mathbf{Z}$, it follows that $ua^2 - 2va \geq 0$.

(c) We use (7.7) with $\zeta_0 = \zeta$ and $\zeta_1 = \operatorname{Ind}_H^L 1_H$. Then $c_1 = (\beta, \chi)$ and $c_i = 0$ for $i \geq 2$. Thus, for $x \in A$,

$$\chi^\rho(x) = c_1 \frac{\zeta_1(x)}{\|\zeta_1\|^2} = (\beta, \chi) \quad \text{and} \quad \|\chi^\rho\|^2 = \frac{|A|}{|L|}(\beta, \chi)^2.$$

$\square$

(7.9) *Assume Hypothesis (7.4) with $I = \{1,2\}$ and that G is of odd order. For $i \in I$, let H_i be a normal subgroup of L_i such that $A_i = H_i^{\#}$ and suppose that $\mathcal{S}_i = \{\operatorname{Ind}_{H_i}^{L_i} \theta \mid \theta \in \operatorname{Irr} H_i,\ \theta \neq 1_{H_i}\}$ is coherent. Let ν_i be an extension of τ_i to an isometry from $\mathbf{Z}[\mathcal{S}_i]$ to $\mathbf{Z}[\operatorname{Irr} G]$. Let $\zeta_i \in \mathcal{S}_i \cap \operatorname{Irr} L_i$ be such that $\zeta_i(1) = |L_i : H_i|$, and let $\beta_i = (\operatorname{Ind}_{H_i}^{L_i} 1_{H_i} - \zeta_i)^{\tau_i}$. Then $(\beta_1, \zeta_2^{\nu_2}) \neq 0$ or $(\beta_2, \zeta_1^{\nu_1}) \neq 0$.*

Proof. Set $\beta_i = 1_G - \zeta_i^{\nu_i} + \Delta_i$. Then $\beta_i - \overline{\beta_i} = (\overline{\zeta_i} - \zeta_i)^{\tau_i} = \overline{\zeta_i^{\nu_i}} - \zeta_i^{\nu_i}$, by (5.9), and so $\overline{\Delta_i} = \Delta_i$. By (7.8.a), $(\Delta_i, 1_G) = 0$. By (1.1), it follows that $(\Delta_1, \Delta_2) = \sum_{\chi \in \operatorname{Irr} G}(\Delta_1, \chi)(\Delta_2, \chi)$ is even. Since $A_1^{\tau_1} \cap A_2^{\tau_2} = \emptyset$, we have $((\zeta_1 - \overline{\zeta_1})^{\tau_1}, (\zeta_2 - \overline{\zeta_2})^{\tau_2}) = (\beta_1, \beta_2) = 0$. By (4.1), $(\zeta_1^{\nu_1}, \zeta_2^{\nu_2}) = 0$. Thus,

$$0 = (\beta_1, \beta_2) = 1 - (\zeta_1^{\nu_1}, \Delta_2) - (\zeta_2^{\nu_2}, \Delta_1) + (\Delta_1, \Delta_2),$$

and so $(\zeta_1^{\nu_1}, \Delta_2) + (\zeta_2^{\nu_2}, \Delta_1) \equiv 1 \pmod{2}$, which proves (7.9). $\qquad\square$

(7.10) *Let G be a group of odd order and k an integer ≥ 2. Assume that:*

(a) *For $1 \leq i \leq k$, L_i is a subgroup of G, L_i is a Frobenius group with kernel H_i, $|H_i| = h_i$ and $|L_i| = h_i e_i$.*

(b) *For all i, $H_i^{\#}$ is a TI-subset of G with normalizer L_i.*

(c) *For $i \neq j$, h_i is prime to h_j.*

(d) $G_0 = G - \bigcup_{1 \leq i \leq k}(H_i^{\#})^G$.

Then there is an index i such that, for $e = e_i$ and $h = h_i$,

$$\frac{|G_0| - 1}{|G|} \geq (e - 1)\left(\frac{h - 2e - 1}{eh} + \frac{2}{h(h + 2)}\right).$$

Proof. The hypothesis implies that Hypothesis (7.4) holds with $A_i = H_i^{\#}$, $\tau_i = \operatorname{Ind}_{L_i}^G$ and $\rho_i = \operatorname{Res}_{L_i}^G$. Let

$$\mathcal{S}_i = \{\operatorname{Ind}_{H_i}^{L_i} \theta \mid \theta \in \operatorname{Irr} H_i,\ \theta \neq 1_{H_i}\} = \{\zeta_{it} \mid 1 \leq t \leq w_i\}.$$

Since L_i is a Frobenius group, $\mathcal{S}_i \subset \operatorname{Irr} L_i$. By a theorem of Thompson, H_i is nilpotent. We may then assume that $\zeta_{i1}(1) = e_i$. Let $\zeta_{it}(1) = d_{it}e_i$. By Theorem (6.8), there are orthonormal subsets $\mathcal{X}_i = \{\chi_{it} \mid 1 \leq t \leq w_i\}$ of $\mathbf{Z}[\operatorname{Irr} G]$ such that

$$\operatorname{Ind}_{L_i}^G(\zeta_{it} - d_{it}\zeta_{i1}) = \chi_{it} - d_{it}\chi_{i1}$$

for $2 \leq t \leq w_i$. Let $i, j \leq k$ be such that $i \neq j$, $1 \leq t \leq w_i$, $1 \leq u \leq w_j$, $\overline{\zeta_{it}} = \zeta_{it'}$ and $\overline{\zeta_{ju}} = \zeta_{ju'}$. As the supports of $\chi_{it} - \chi_{it'}$ and of $\chi_{ju} - \chi_{ju'}$ are disjoint, $(\chi_{it} - \chi_{it'}, \chi_{ju} - \chi_{ju'}) = 0$. By (4.1), it follows that $\mathcal{X}_i$ is orthogonal to $\mathcal{X}_j$. By (7.8.a), $\mathcal{X}_i$ is orthogonal to 1_G.

Without loss of generality we will assume that the index 1 is such that $h_1 \leq h_i$ for $1 \leq i \leq k$. Set $L = L_1$, $H = H_1$, $h = h_1$, $e = e_1$, $\chi_t = \chi_{1t}$, $\zeta_t = \zeta_{1t}$, $d_t = d_{1t}$, $w = w_1$ and $\rho = \rho_1$. Let $\beta_i = \operatorname{Ind}_{L_i}^G(\operatorname{Ind}_{H_i}^{L_i} 1_{H_i} - \zeta_{i1})$ and $\beta = \beta_1$. By (7.8.a),

$$\beta = 1_G - \chi_1 + a\sum_{t=1}^{w} d_t \chi_t + \Gamma$$

where $a \in \mathbf{Z}$ and, consequently, $\Gamma \in \mathbf{Z}[\mathrm{Irr}\, G]$, and Γ is orthogonal to $\mathcal{X}_1 \cup \{1_G\}$. Let $\mathcal{A}$ be the set of indices $i > 1$ such that $(\beta_i, \chi_1) \neq 0$. Let $\mathcal{B}$ be the set of indices $i > 1$ which do not belong to $\mathcal{A}$. By (7.5),

$$
\begin{aligned}
\frac{|G_0| - 1}{|G|} \;&\geq\; \frac{|G_0| - \chi_1(1)^2}{|G|} \\[2mm]
&\geq\; \sum_{1 \leq i \leq k} \left(\|\chi_1^{\rho_i}\|^2 - \frac{|A_i|}{|L_i|} \right) \\[2mm]
&\geq\; \|\chi_1^\rho\|^2 - \frac{|H^\#|}{|L|} + \sum_{i \in \mathcal{A}} \left(\|\chi_1^{\rho_i}\|^2 - \frac{|H_i^\#|}{|L_i|} \right) - \sum_{i \in \mathcal{B}} \frac{|H_i^\#|}{|L_i|}.
\end{aligned}
$$

Since L is a Frobenius group, e divides $h - 1$. As $|L|$ is odd, $e \leq (h - 1)/2$. By (7.8.b), $\|\chi_1^\rho\|^2 \geq 1 - e/h$. By (7.8.c), $\|\chi_1^{\rho_i}\|^2 \geq |H_i^\#|/|L_i|$ for $i \in \mathcal{A}$. Then

$$
\frac{|G_0| - 1}{|G|} \geq 1 - \frac{e}{h} - \frac{h - 1}{eh} - \sum_{i \in \mathcal{B}} \frac{h_i - 1}{e_i h_i}.
$$

For $i > 1$, β and $\chi_{it} - d_{it}\chi_{i1}$ have disjoint supports, and so $(\beta, \chi_{it} - d_{it}\chi_{i1}) = 0$. Thus there are integers x_i such that $\Gamma = \sum_{i>1} x_i \sum_t d_{it}\chi_{it} + \Gamma_1$, where Γ_1 is orthogonal to $\bigcup_i \mathcal{X}_i$. By (7.9), we have $x_i \neq 0$ for $i \in \mathcal{B}$. By (7.8.b), it follows that $\sum_{i \in \mathcal{B}} \sum_t d_{it}^2 \leq e - 1$. But $\sum_t d_{it}^2 = (|L_i| - |L_i : H_i|)/e_i^2 = (h_i - 1)/e_i$. Thus

$$
\sum_{i \in \mathcal{B}} \frac{h_i - 1}{h_i e_i} \leq \frac{1}{h + 2} \sum_{i \in \mathcal{B}} \frac{h_i - 1}{e_i} \leq \frac{e - 1}{h + 2}.
$$

We thus obtain

$$
\begin{aligned}
\frac{|G_0| - 1}{|G|} \;&\geq\; 1 - \frac{e}{h} - \frac{h - 1}{eh} - \frac{e - 1}{h} + (e - 1)\left(\frac{1}{h} - \frac{1}{h + 2} \right) \\[2mm]
&=\; (e - 1)\left(\frac{h - 2e - 1}{eh} + \frac{2}{h(h + 2)} \right).
\end{aligned}
$$

$\square$

(7.11) Theorem. *There is no group satisfying the hypothesis of (7.10) with* $G_0 = \{1\}$.

Proof. Suppose that the hypothesis of (7.10) holds with $|G_0| = 1$. Since e divides $h - 1$ and h is odd, $\dfrac{h - 2e - 1}{eh} \geq 0$. Thus,

$$
(e - 1)\left(\frac{h - 2e - 1}{eh} + \frac{2}{h(h + 2)} \right) > 0,
$$

which contradicts (7.10). $\square$

8. Structure of a Minimal Simple Group of Odd Order

Assume throughout that G is a non-abelian simple group of odd order in which every proper subgroup is solvable. We say that such a group is a *minimal simple group of odd order*. We will show that such a group does not exist. For this, we rely upon the results set out in [FT], Chapter IV, and in [BG], which describe the structure of the maximal subgroups of G.

If M is a finite group, we denote by M_F the unique normal nilpotent Hall subgroup of M which is maximal with these properties.

(8.1) Definition. Let M be a solvable group of odd order and let $H = M_F$. We say that M is *of type $\mathcal{F}$* if:

(a) $1 \neq H \subset M$ and there is a subgroup $U \neq 1$ of M such that $M = H \rtimes U$.

(b) There is a normal abelian subgroup U_1 of U such that $C_U(x) \subset U_1$ for all $x \in H^{\#}$.

(c) There is a subgroup U_0 of U, which has the same exponent as U, such that HU_0 is a Frobenius group with kernel H.

With this notation, every complement of H in M is conjugate to U in M by a theorem of Hall. Thus, (b) and (c) hold whatever complement U is chosen.

(8.2) *Let M be a group of type $\mathcal{F}$. Using the notation of Definition (8.1), we have:*

(a) $|U_0| = \exp(U)$.

(b) *M is a Frobenius group with kernel H if and only if the Sylow subgroups of U are cyclic.*

(c) *If $\theta \in \operatorname{Irr} H - \{1_H\}$, then $I(\theta) \cap U \subset U_1$.*

Proof. (a) Since U_0 is a Frobenius complement of odd order, the Sylow subgroups of U_0 are cyclic by [BG], Proposition 3.9. Let π be the set of prime divisors of $|U_0|$. If $p \in \pi$, U_0 contains an element of order $|U_0|_p$, and so $|U_0| = \prod_{p \in \pi} |U_0|_p$ divides $\exp(U_0)$. As $\exp(U_0)$ divides $|U_0|$, it follows that $|U_0| = \exp(U_0) = \exp(U)$.

(b) If M is a Frobenius group with kernel H, we know that the Sylow subgroups of U are cyclic. Conversely, if the Sylow subgroups of U are cyclic, $\exp(U) = |U|$. Thus, by (a), $|U_0| = |U|$, whence $U_0 = U$ and M is a Frobenius group with kernel H by (8.1.c).

(c) Let $g \in U - U_1$. Let C be a conjugacy class of H normalized by g. Let $a \in C$. There is an element $x \in H$ such that $a^{g^{-1}} = a^x$. By (2.1), there are elements $y \in H$ and $z \in C_H(g)$ such that $(xg)^y = zg$. Then zg centralizes $a^y \in C$ and, since g is a power of zg, g centralizes an element of C. Thus

$C = \{1\}$. It follows from [Is], Theorem 6.32, that the number of irreducible characters of H invariant under g is 1. Consequently, if $\theta \in \operatorname{Irr} H - \{1_H\}$, then $g \notin I(\theta)$. $\qquad\qquad\square$

(8.3) Definition. Let M be a subgroup of G and $H = M_F$. We say that M is *of Type I* if M is of type $\mathcal{F}$ and, in the notation of Definition (8.1), one of the following conditions is satisfied.

(a) $H^{\#}$ is a TI-subset of G.

(b) H is abelian of rank 2.

(c) For every prime divisor p of $|H|$, the exponent of U divides $p - 1$, and there is a prime divisor p of $|H|$ such that $O_{p'}(M)$ is cyclic.

(8.4) Definition. Let M be a subgroup of G, $H = M_F$, $M' = [M, M]$ and $M'' = [M', M']$. We say that M is *of type $\mathcal{P}$* if it satisfies the following conditions.

(a) There is a cyclic Hall subgroup $W_1 \neq 1$ of M such that $M = M' \rtimes W_1$.

(b) There is a nilpotent subgroup U of M', normalized by W_1, such that $M' = H \rtimes U$.

(c) H is not cyclic and $M'' \subset HC_M(H) = F(M) \subset M'$.

(d) There is a cyclic subgroup $W_2 \neq 1$ of $H \cap M''$ such that $C_{M'}(x) = W_2$ for all $x \in W_1^{\#}$.

(e) Let $W = W_1 W_2$ and $V = W - (W_1 \cup W_2)$. For every non-empty subset X of V, $N_G(X) = W$.

If M is a subgroup of G of type $\mathcal{P}$, then M is solvable. By a theorem of Hall, every complement of M' in M is conjugate to W_1 in M. Therefore, properties (8.4.b–e) hold whatever complement W_1 is chosen.

(8.5) *Let M be a subgroup of G of type $\mathcal{P}$. Using the notation of Definition (8.4), we have:*

(a) $F(M) = HC_U(H)$.

(b) $[U, U]$ *centralizes H. If $U \neq 1$, then U does not centralize H.*

(c) V *is a TI-subset of G with normalizer W.*

Proof. (a) By (8.4.c), $HC_U(H) \subset HC_M(H) = F(M)$. Let $g \in C_M(H)$. By (8.4.c), $g \in M' = HU$. Let $h \in H$ and $u \in U$ be such that $g = hu$. Then, in the group $M/C_M(H)$, $\overline{h} = \overline{u}^{-1}$. But $\overline{h}$ and $\overline{u}$ have relatively prime orders. Thus $\overline{h} = \overline{u} = 1$ and $u \in C_U(H)$. Consequently, $g \in HC_U(H)$ and $F(M) = HC_U(H)$.

(b) By (8.4.b,c), we see that $U \subset M'$ and $[U, U] \subset M'' \subset F(M)$. Also, by (a), $[U, U] \subset HC_U(H)$ and so $[U, U] \subset C_U(H)$. If U centralizes H, then HU is nilpotent, and so $HU = M_F$ and $U = 1$.

(c) Let $g \in G$ be such that $V \cap V^g \neq \emptyset$. Let $v \in V \cap V^g$. Then $v \in W^g$ and W^g is abelian. By (8.4.e), $W^g \subset C_G(v) = W$. Thus g normalizes W. As W_1 and W_2 are characteristic in W, $g \in N_G(V)$ and so $g \in W$ by (8.4.e).   □

(8.6) Definition. Let M be a subgroup of G. We say that M is *of Type II, III* or *IV* if it is of type $\mathcal{P}$ and if, in the notation of Definition (8.4),

(a) $U \neq 1$, $|W_1|$ is prime and $F(M)^\#$ is a TI-subset of G.

Moreover, for the subgroups of Type II,

(b II) U is abelian, $N_G(U) \not\subset M$ and M' is of type $\mathcal{F}$, with $(M')_F = H$.

For the subgroups of Type III,

(b III) U is abelian and $N_G(U) \subset M$.

For the subgroups of Type IV,

(b IV) U is not abelian and $N_G(U) \subset M$.

(8.7) Definition. Let M be a subgroup of G. We say that M is *of Type V* if it is of type $\mathcal{P}$ and if, in the notation of Definition (8.4), $U = 1$ and one of the following conditions is satisfied:

(a) $H^\#$ is a TI-subset of G.

(b) There is a prime divisor p of $|H|$ such that $|W_1|$ divides $p - 1$ and $O_{p'}(H)$ is cyclic.

(c) There is a prime divisor p of $|H|$ such that $|O_p(H)| = p^3$, $|W_1|$ divides $p + 1$ and $O_{p'}(H)$ is cyclic.

(8.8) Theorem. *One of the following two cases holds.*

(a) *Every maximal subgroup of G is of Type I.*

(b) *G contains a cyclic subgroup $W = W_1 \times W_2$, with $W_1 \neq 1$ and $W_2 \neq 1$, which satisfies (8.4.e), and two maximal subgroups S and T such that:*

(b1) $S = [S, S] \rtimes W_1$, $T = [T, T] \rtimes W_2$ *and* $S \cap T = W$.

(b2) S *or* T *is of Type II.*

(b3) S *and* T *are of Type II, III, IV or V.*

(b4) *Every maximal subgroup of G is conjugate to S or to T, or is of Type I.*

Reference. [BG], § 16, Theorem I, Proposition 16.1, Theorem B and Theorem C(3).

(8.9) *Suppose that case* (b) *of Theorem* (8.8) *holds. Then the group denoted by W_2 in Theorem* (8.8) *coincides with the group denoted by W_2 in* (8.4.d) *with $M = S$.*

Proof. In the notation of Theorem (8.8), $W_2 \subset W \subset S$. Since W is cyclic, $|W_1|$ and $|W_2|$ are relatively prime, and so $W_2 \subset S'$, the commutator subgroup of S. Thus $W_2 \subset C_{S'}(W_1)$. By (8.4.d) with $M = S$, $W_1 C_{S'}(W_1)$ is abelian, and so $C_{S'}(W_1) \subset C(W)$. As W satisfies (8.4.e), $C_{S'}(W_1) \subset W$, whence $C_{S'}(W_1) = W_2$. $\qquad\qquad\square$

(8.10) Notation. Let M be a maximal subgroup of G and let $H = M_F$. If M is of type $\mathcal{P}$, we use the notation of Definition (8.4). Set

$$M_s = \begin{cases} H & \text{if } M \text{ is of Type I, II or V,} \\ M' & \text{if } M \text{ is of Type III or IV.} \end{cases}$$

Set $A_1(M) = M_s^\#$. If M is of Type I, set $A(M) = A_0(M) = \bigcup_{x \in H^\#} C_M(x)^\#$. If M is of type $\mathcal{P}$, set $A(M) = \bigcup_{x \in M_s^\#} C_{M'}(x)^\#$ and $A_0(M) = A(M) \cup V^M$.

Thus, $A_1(M) \subset A(M) \subset A_0(M)$ and $A_1(M) = A(M) = (M')^\#$ if M is of Type III, IV or V.

(8.11) *Let M be a maximal subgroup of G. If P is a non-trivial Sylow subgroup of M_s, then $N_G(P) \subset M$. Moreover, M_F and M_s are Hall subgroups of G.*

Reference. By [BG], Proposition 16.1, M_s is the group denoted by M_σ in [BG], and (8.11) follows from [BG], §16, Theorem A(1).

(8.12) *Let M be a maximal subgroup of G of Type I or II and let $H = M_F$. If M is of Type I, let $M = H \rtimes U$, and, if M is of Type II, let $[M, M] = H \rtimes U$.*

(a) *Every Sylow subgroup of U is abelian of rank ≤ 2.*

(b) *For every non-empty subset X of $U^\#$ such that $C_H(X) \neq 1$, M is the unique maximal subgroup of G which contains $C_G(X)$.*

(c) *$A(M) - A_1(M)$ is a TI-subset of G.*

Reference. [BG], §16, Theorem B and Proposition 16.1.

(8.13) Theorem. *Let M be a maximal subgroup of G, and $X = A(M)$ or $X = A_0(M)$. Let $D = \{x \in X \mid C_G(x) \not\subset M\}$.*

(a) *If two elements of X are conjugate in G, they are conjugate in M.*

(b) *$D \subset A_1(M)$ and, for $x \in D$, $C_G(x)$ is contained in a unique maximal subgroup of G.*

(c) *For $x \in D$ and L the maximal subgroup of G for which $C_G(x) \subset L$:*

(c1) *$L = L_F \rtimes (M \cap L)$ and $C_G(x) = C_{L_F}(x) \rtimes C_M(x)$.*

(c2) *$|L_F|$ is prime to $|C_M(y)|$ for all $y \in X$.*

(c3) *$x \in A(L) - A_1(L)$.*

(c4) *L is of Type I or II. Furthermore, M is a Frobenius group with kernel M_F if L is of Type II.*

Reference. [BG], §16, Theorem II, Theorem B(5) and Theorem D(4).

(8.14) Definition and Notation. In the notation of Theorem (8.13), let L be a maximal subgroup of G such that there is an element $x \in D$ with $C_G(x) \subset L$. We then say that L *supports* M. If $x \in A_0(M) - D$, set $R(x) = 1$. If $x \in D$ and L is the maximal subgroup of G for which $C_G(x) \subset L$, set $R(x) = C_{L_F}(x)$. Set

$$\tilde{A}(M) \quad = \quad \bigcup_{a \in A(M)} (aR(a))^G,$$

$$\widetilde{A_0}(M) \quad = \quad \bigcup_{a \in A_0(M)} (aR(a))^G$$

and

$$\widetilde{A_1}(M) \quad = \quad \bigcup_{a \in A_1(M)} (aR(a))^G.$$

(8.15) *Let M be a maximal subgroup of G and let $A = A_0(M)$, $A(M)$ or $A_1(M)$. Then $M = N_G(A)$ and Hypothesis (2.2) holds with the groups denoted by L and $H(a)$ in Hypothesis (2.2) being M and $R(a)$ respectively. If M is of type $\mathcal{P}$ and $M' = [M, M]$, then Hypothesis (4.6) holds for*
$$L = M, \; K = M', \; A = A(M), \; A_0 = A_0(M)$$
and
$$H = M_F \quad or \quad H = M_s.$$
If M is of type $\mathcal{P}$ and $\mathcal{S}$ is a non-empty subset of
$$\{\mathrm{Ind}_{M'}^M \theta \mid \theta \in \mathrm{Irr}\, M', \; M_s \not\subset \mathrm{Ker}\, \theta\}$$
closed under complex conjugation, then Hypothesis (5.2) holds for $L = M$.

Proof. Clearly, $A \subset M \subset N_G(A)$. If $N_G(A) = G$, then $\langle A \rangle$ is a normal subgroup of G distinct from 1 and from G, which contradicts the simplicity of G. Since M is maximal, $M = N_G(A)$. Statements (2.2.a, b, c) hold by (8.13.a, c1, c2). The second assertion then follows from (8.4.a, d), (8.5.c) and (8.10), and the third follows from (1.5.e) and (5.3.b). $\qquad\qquad\square$

(8.16) *Let M be a maximal subgroup of G of Type II. Then $A_0(M)$, $A(M)$ and $A_1(M)$ are TI-subsets of G with normalizer M.*

Proof. If $a \in A_0(M) - A_1(M)$, then $R(a) = 1$ by (8.13.b). If $a \in A_1(M)$, then (8.6.a) implies that $R(a) = 1$. The conclusion then follows from (2.3). $\square$

(8.17) Theorem. *Let $M_1, \ldots, M_n$ be a system of representatives of the conjugacy classes of maximal subgroups of G.*

   (a) *$\pi(G)$ is the disjoint union of its subsets $\pi((M_i)_s)$, $1 \le i \le n$.*

   (b) *$|\widetilde{A_1}(M_i)| = (|(M_i)_s| - 1)|G : M_i|$ for $1 \le i \le n$.*

   (c) *In case (a) of Theorem (8.8), $G^{\#}$ is the disjoint union of its subsets $\widetilde{A_1}(M_i)$; in case (b), $G^{\#}$ is the disjoint union of its subsets $\widetilde{A_1}(M_i)$ and of V^G.*

Reference. [BG], §16, Theorem E.

(8.18) *Let S and T be two non-conjugate maximal subgroups of G.*

(a) *T supports S if and only if $A_1(S) \cap A(T) \neq \emptyset$. If $x \in A_1(S) \cap A(T)$, then $C_G(x) \not\subset S$ and $C_G(x) \subset T$.*

(b) *There is a conjugate of T supporting S if and only if $\widetilde{A}_1(S) \cap \widetilde{A}(T) \neq \emptyset$.*

(c) *$\widetilde{A}_1(S) \cap \widetilde{A}(T) = \emptyset$ or $\widetilde{A}_1(T) \cap \widetilde{A}(S) = \emptyset$.*

Proof. (a) By $(8.13.\mathrm{b}, \mathrm{c}3)$, $A_1(S) \cap A(T) \neq \emptyset$ if T supports S. Conversely, let $x \in A_1(S) \cap A(T)$. By $(8.17.\mathrm{a})$, x has order prime to $|T_s|$ and $x \notin A_1(T)$. Since $A(T) - A_1(T) \neq \emptyset$, T is of Type I or II. Since x has order prime to $|T_s|$, it follows from $(8.12.\mathrm{b})$ that T is the unique maximal subgroup of G such that $C_G(x) \subset T$. Thus $C_G(x) \not\subset S$ and T supports S.

(b) If a conjugate T^g of T supports S, then, by (a), $A_1(S) \cap A(T)^g \neq \emptyset$ and therefore $\widetilde{A}_1(S) \cap \widetilde{A}(T) \neq \emptyset$. Suppose conversely that $\widetilde{A}_1(S) \cap \widetilde{A}(T) \neq \emptyset$. Replacing S and T by conjugate subgroups, we may assume that there are elements $a \in A(T)$ and $b \in A_1(S)$ such that $aR(a) \cap bR(b) \neq \emptyset$. But, by $(8.17.\mathrm{c})$, $a \in A(T) - A_1(T)$, and so $R(a) = 1$. There is thus an element $x \in R(b)$ such that $a = bx$. Then b is a power of a, and so $b \in A(T)$ by the definition of $A(T)$. Thus, $A_1(S) \cap A(T) \neq \emptyset$ and T supports S by (a).

(c) Suppose that $\widetilde{A}_1(S) \cap \widetilde{A}(T) \neq \emptyset$. By (b), we may assume that T supports S. By $(8.13.\mathrm{c}2, \mathrm{c}4)$, $|T_s|$ is then prime to $|C_S(a)|$ for all $a \in A_0(S)$. Thus $A_1(T) \cap A(S)^g = \emptyset$ for all $g \in G$. By (a), no conjugate of S supports T, and so, by (b), $\widetilde{A}_1(T) \cap \widetilde{A}(S) = \emptyset$. $\qquad\square$

9. On the Maximal Subgroups of G of Types II, III and IV

(9.1) *Let $U \rtimes E$ be a Frobenius group with kernel U. Assume that UE acts on a finite solvable group H of order prime to $|UE|$. Then*

$$|C_H(UE)|^{|E|}|H| = |C_H(E)|^{|E|}|C_H(U)|.$$

In particular, if $C_H(E) = 1$, then U centralizes H, and, if $C_H(U) = 1$, then $|H| = |C_H(E)|^{|E|}$.

Proof. Since H is assumed to be solvable, Wielandt's fixed point theorem ([HB], Chapter XI, Theorem 12.4) holds for the action of UE on H, without assuming the Feit-Thompson Theorem. If X is a subset of UE, we denote by $\underline{X}$ the sum of the elements of X taken in the algebra $\mathbf{Z}[UE]$. Since UE is a Frobenius group,

$$\underline{UE} + |U|\underline{1} = \textstyle\sum_{u \in U} \underline{E^u} + \underline{U}.$$

Thus, by the theorem cited,

$$|C_H(UE)|^{|UE|}|H|^{|U|} = \left(\prod_{u \in U} |C_H(E^u)|^{|E|} \right)|C_H(U)|^{|U|},$$

and the first assertion follows. The special cases follow because

$$C_H(UE) \subset C_H(U) \cap C_H(E).$$

$\square$

(9.2) Hypothesis. *M is a maximal subgroup of G of Type II, III or IV while H, U, W_1 and W_2 have the same meaning as in Definition (8.4) and $q = |W_1|$.*

(9.3) *Assume Hypothesis (9.2). If M is of Type II, then $C_H(U) = 1$ and $|H| = |W_2|^q$. If M is of Type III or IV, then $p = |W_2|$ is prime, $C_H(UW_1) = 1$ and $|H| = p^q|C_H(U)|$.*

Proof. By Definition (8.4), UW_1 is a Frobenius group with kernel U, which acts on H. Suppose that M is of Type II. If $C_H(U) \neq 1$, then, by (8.6.b II) and (8.12.b), $N_G(U) \not\subset M$ as well as $N_G(U) \subset M$. Thus $C_H(U) = 1$, and $|H| = |C_H(W_1)|^q = |W_2|^q$ by (9.1). Suppose that M is of Type III or IV. By Theorem (8.8), there is a maximal subgroup S of Type II such that $|S : [S, S]| = |W_2|$. Thus $p = |W_2|$ is prime. By (9.1), $|C_H(UW_1)|^q|H| = p^q|C_H(U)|$. Suppose that $C_H(UW_1) \neq 1$. Since $C_H(UW_1) \subset C_H(W_1) = W_2$, $|C_H(UW_1)| = p$ and $|H| = |C_H(U)|$. This contradicts (8.5.b), and so we see that $C_H(UW_1) = 1$ and $|H| = p^q|C_H(U)|$. $\square$

(9.4) *Assume Hypothesis (9.2). Then there is a normal subgroup H_0 of M and a prime number p such that:*

(a) $H_0 \subset H$ and $\overline{H} = H/H_0$ is a non-trivial elementary abelian p-group.

(b) If M is of Type III or IV, then $p = |W_2|$ and $\overline{H}$ is a chief factor of M not centralized by U.

Proof. This is clear if M is of Type II because H is nilpotent, $H \neq 1$. Suppose that M is of Type III or IV. Let $p = |W_2|$ and let P be the Sylow p-subgroup of H. By Maschke's Theorem, $P/\Phi(P) = P_1 \times \cdots \times P_r$ where the P_i are subgroups of $P/\Phi(P)$ normalized by UW_1 such that UW_1 acts irreducibly on each P_i. By (9.3), $|H : C_H(U)| = p^q$ and so U does not centralize P. By [BG], Theorem 1.8, U does not centralize $P/\Phi(P)$. There is thus an index i such that U does not centralize P_i. It suffices to take as H_0 the product of $O_{p'}(H)$ and the inverse images in P of the subgroups P_j for $j \neq i$. $\qquad\square$

In the remainder of this section, we will assume the following hypothesis.

(9.5) Hypothesis. *Assume that Hypothesis* (9.2) *holds. Let H_0 be as in* (9.4). *Set $\overline{H} = H/H_0$, $C = C_U(\overline{H})$, $\overline{U} = U/C$, $u = |\overline{U}|$, $\overline{W}_2 = C_{\overline{H}}(W_1)$, $U' = [U,U]$ and $C' = [C,C]$. Let τ denote the Dade isometry relative to $(A(M), M, G)$. Set $\mathcal{X} = \{\chi \in \mathrm{Irr}\, HU \mid H \not\subset \mathrm{Ker}\,\chi\}$ and set $\mathcal{S} = \{\mathrm{Ind}_{HU}^M \chi \mid \chi \in \mathcal{X}\}$. In addition, for $Y \subset HU$, set $\mathcal{X}(Y) = \{\chi \in \mathcal{X} \mid Y \subset \mathrm{Ker}\,\chi\}$ and set $\mathcal{S}(Y) = \{\mathrm{Ind}_{HU}^M \chi \mid \chi \in \mathcal{X}(Y)\}$.*

(9.6) $U \neq C$, $\overline{H}$ *is a chief factor of M,* $|\overline{W}_2| = p$ *and* $|\overline{H}| = p^q$.

Proof. The first two assertions are true by hypothesis if M is of Type III or IV. Suppose that M is of Type II. By (9.3) and [BG], Proposition 1.5(d), $C_{\overline{H}}(U) = 1$, and so $U \neq C$. Suppose that $\overline{H}$ is not a chief factor of M. By Maschke's Theorem, there are non-trivial subgroups $\overline{H}_1$ and $\overline{H}_2$ of $\overline{H}$ normalized by UW_1 such that $\overline{H} = \overline{H}_1 \times \overline{H}_2$. By [BG], Proposition 1.5(d), $\overline{W}_2$ is the image of W_2 in $\overline{H}$, and so is cyclic. We may then assume that $C_{\overline{H}_1}(W_1) = 1$. By (9.1), it follows that U centralizes $\overline{H}_1$, which is a contradiction because $C_{\overline{H}}(U) = 1$. Thus $\overline{H}$ is a chief factor of M.

The subgroup $C_{\overline{H}}(U)$ of $\overline{H}$ is normalized by UW_1 and $C_{\overline{H}}(U) \neq \overline{H}$. It follows that $C_{\overline{H}}(U) = 1$. Then $|\overline{H}| = |\overline{W}_2|^q$ by (9.1). As $\overline{W}_2$ is cyclic, $|\overline{W}_2|$ divides p, whence $|\overline{W}_2| = p$ and $|\overline{H}| = p^q$. $\qquad\square$

(9.7) *One of the following two cases holds.*

(a) $\overline{H}$ *is the direct product of q groups H_i $(1 \leq i \leq q)$ of order p normalized by U; moreover, $\{H_i \mid 1 \leq i \leq q\} = \{H_1^w \mid w \in W_1\}$. Let $a = |U : C_U(H_1)|$. Then a divides $p - 1$, $U/C_U(H_i)$ is cyclic of order a for all i and $\overline{U}$ is isomorphic to a subgroup of the direct product of $q - 1$ cyclic groups of order a.*

(b) U *acts irreducibly on $\overline{H}$. There is a field F of cardinality p^q and a subgroup U^* of F^* such that $(\overline{H} \rtimes \overline{U}) \rtimes W_1$ is isomorphic to $(F \rtimes U^*) \rtimes (\mathrm{Aut}\, F)$, where U^* acts on the additive group F by multiplication and $\mathrm{Aut}\, F$ acts naturally on F and U^*. This isomorphism identifies $\overline{H}$ ($\overline{U}$ and W_1, respectively)* *with the additive group F (with U^* and with $\operatorname{Aut} F$, respectively). Furthermore, $\overline{U}$ is cyclic, and u is prime to $p-1$ and divides $(p^q - 1)/(p - 1)$.*

Proof. We use additive notation for the operation in $\overline{H}$. By (9.6), UW_1 acts irreducibly on $\overline{H}$. By Clifford's Theorem ([Is], Theorem 6.5),
$$\overline{H} = H_1 \oplus \cdots \oplus H_k$$
where the H_i are irreducible $\mathbf{F}_p[U]$-modules and are conjugate under W_1. Thus $q = \dim(\overline{H}) = k \dim(H_1)$. Since q is prime, $k = q$ or $k = 1$.

Suppose that $k = q$. Then $|H_1| = p$ and all the conjugates of H_1 under W_1 are among the modules H_i. Let w be a generator of W_1. We may assume that $H_{i+1} = H_i^w$ for $1 \leq i \leq q$, with $H_{q+1} = H_1$. Since $\operatorname{Aut}(\mathbf{Z}/p\mathbf{Z}) \cong (\mathbf{Z}/p\mathbf{Z})^*$, there are homomorphisms $\varphi_i : U \to (\mathbf{Z}/p\mathbf{Z})^*$ for which $h^x = \varphi_i(x)h$ for $h \in H_i$ and $x \in U$. Then $h^{wx} = h^{wxw^{-1}w} = (\varphi_i(wxw^{-1})h)^w$ for $h \in H_i$ and $x \in U$. Thus $\varphi_{i+1}(x) = \varphi_i(wxw^{-1})$. The group $U/C_U(H_i)$ is isomorphic to the image of φ_i, and so is cyclic of order dividing $p - 1$. As H_i is conjugate to H_1 under W_1, $|U/C_U(H_i)| = a$. Let φ be the homomorphism from U to $((\mathbf{Z}/p\mathbf{Z})^*)^{q-1}$ for which $\varphi(x) = (\varphi_2(x)\varphi_1(x)^{-1}, \ldots, \varphi_q(x)\varphi_1(x)^{-1})$ for $x \in U$. If $x \in \operatorname{Ker}\varphi$, then $\varphi_i(x) = \varphi_1(x)$ for all i, and so x and wxw^{-1} act in the same way on $\overline{H}$. As $\overline{U} \rtimes W_1$ is a Frobenius group, it follows that $x \in C$. Thus $\overline{U}$ is isomorphic to $\operatorname{Im}\varphi$, which is a subgroup of the product of $q - 1$ cyclic groups of order a.

Suppose that $k = 1$. By (8.5.b), U' centralizes H, and so $\overline{U}$ is abelian. Let F be the ring of endomorphisms of the $\mathbf{F}_p[U]$-module $\overline{H}$. For α, $\beta \in F$ and $h \in \overline{H}$, we write $h\alpha$ for the image of h under α and $h(\alpha\beta) = (h\alpha)\beta$. By Schur's Lemma ([Is], Lemma 1.5), F is a finite field. Since $\overline{U}$ is abelian, there is an isomorphism ψ from $\overline{U}$ onto a subgroup U^* of F^* such that $h\psi(x) = h^x$ for $h \in \overline{H}$. Since U acts on $\overline{H}$ irreducibly, $\overline{H}$ is a vector space over F of dimension 1 and the additive subgroup of F generated by U^* is F. By (9.6), there is an element $s \in \overline{W_2}^{\#}$. Let φ be the additive isomorphism from $\overline{H}$ to F for which $h = s\varphi(h)$ for $h \in \overline{H}$. If $h \in \overline{H}$ and $x \in \overline{U}$, then $s\varphi(h)\psi(x) = h^x = s\varphi(h^x)$, and so $\varphi(h^x) = \varphi(h)\psi(x)$. For $w \in W_1$, let $\eta(w)$ be the mapping from F to itself for which $\varphi(h)\eta(w) = \varphi(h^w)$ for $h \in \overline{H}$. Then η is an injective homomorphism from W_1 to the group of additive automorphisms of F. For $h \in \overline{H}$, $x \in \overline{U}$ and $w \in W_1$, $h^{xw} = (h^w)^{w^{-1}xw}$, whence

$$\varphi(h^x)\eta(w) = \varphi(h^w)\psi(w^{-1}xw)$$

and

$$(\varphi(h)\psi(x))\eta(w) = (\varphi(h)\eta(w))\psi(w^{-1}xw).$$

In particular, if $h = s$, we obtain $\psi(x)\eta(w) = \psi(w^{-1}xw)$. Thus, for all h, $(\varphi(h)\psi(x))\eta(w) = (\varphi(h)\eta(w))(\psi(x)\eta(w))$. Since U^* generates the additive group F, $\eta(w)$ is thus an automorphism of the field F. Since U^* is a subgroup of F^*, U^* is cyclic and u divides $p^q - 1$. As W_1 acts fixed-point-freely on $\overline{U}$, $U^* \cap \mathbf{F}_p = 1$. Thus u is prime to $p - 1$ and divides $(p^q - 1)/(p - 1)$. $\qquad\square$

(9.8) *Suppose that case (9.7.a) holds.*

  (a) *If $\chi \in \mathcal{X}(H_0)$, then $\chi(1)$ is divisible by a.*

  (b) *$\mathcal{S}(H_0)$ contains exactly $p - 1$ reducible characters μ_j $(1 \leq j < p)$. For $1 \leq j < p$, $\mu_j(1) = qu$, $\mu_j \in \mathcal{S}(H_0C)$ and μ_j is induced to M from a linear character of HC.*

  (c) *$\mathcal{S}(H_0C)$ contains an irreducible character of degree qu induced from a linear character of HC.*

  (d) *$\mathcal{S}(H_0U')$ contains at least $\dfrac{p-1}{a}\dfrac{|U|}{a|U'|}$ irreducible characters which are of degree qa.*

Proof. For normal subgroups X and Y of M such that $Y \subset X$, we identify the characters of X having Y in their kernels with characters of X/Y. By (1.6), this identification commutes with induction.

(a) Let $\theta = \theta_1 \ldots \theta_q$ be an irreducible character of $\overline{H}$, where $\theta_i \in \mathrm{Irr}\, H_i$. Let J be the set of indices i such that $\theta_i \neq 1$. Since U normalizes the subgroups H_i, $I(\theta) \cap U = \bigcap_{i \in J}(I(\theta_i) \cap U)$. For $i \in J$, θ_i is injective, and so $I(\theta_i) \cap U = C_U(H_i)$. Thus $I(\theta) \cap U = \bigcap_{i \in J} C_U(H_i)$. If $\theta \neq 1$, then $J \neq \emptyset$ and there is an index i such that $I(\theta) \cap U \subset C_U(H_i)$. Thus $|U : I(\theta) \cap U|$ is divisible by $a = |U : C_U(H_i)|$. Now let θ be an irreducible component of $\mathrm{Res}^{HU}_H \chi$. Since $H \not\subset \mathrm{Ker}\,\chi$, $\theta \neq 1$. By (1.7.a), χ is induced to HU from an irreducible character of $I(\theta) \cap HU$, and so $\chi(1)$ is divisible by $|U : I(\theta) \cap U|$, and so by a.

(b,c) By (8.4.d), Hypothesis (4.2) holds for $L = M/H_0$ and also for $L = M/H_0C$ with the group denoted by W_2 in Hypothesis (4.2) being $\overline{W}_2$. By (4.7) and Theorem (4.5), $\mathcal{S}(H_0)$ and $\mathcal{S}(H_0C)$ contain exactly $p - 1$ reducible characters μ_j $(1 \leq j < p)$. Let $\theta = \theta_1 \ldots \theta_q 1_C \in \mathrm{Irr}(HC)$, where $\theta_i \in \mathrm{Irr}\, H_i$, $\theta_i \neq 1$. Then $I(\theta) \cap U = \bigcap_{1 \leq i \leq q} C_U(H_i) = C$. Thus $\chi = \mathrm{Ind}^{HU}_{HC}\theta$ is irreducible. As each such character θ has u conjugates under U, we thus obtain $(p-1)^q/u$ elements χ of $\mathcal{X}(H_0C)$ such that $\chi(1) = u$.

Let $\theta_1 \in \mathrm{Irr}\, H_1$, $\theta_1 \neq 1_{H_1}$. Let $\theta_i \in \mathrm{Irr}\, H_i$ $(1 \leq i \leq q)$ be such that, in the notation of the proof of (9.7), $\theta_{i+1}(h^w) = \theta_i(h)$ for $h \in H_i$. Let $\theta = \theta_1 \ldots \theta_q 1_C$ and $\chi = \mathrm{Ind}^{HU}_{HC}\theta$. Then θ and χ are invariant under W_1, and so $\mathrm{Ind}^M_{HU}\chi$ is one of the characters μ_j. If θ' is a character of HC constructed in the same way and if $x \in U$ is such that $\theta' = \theta^x$, then $\theta^{wxw^{-1}} = \theta^{xw^{-1}} = \theta'^{w^{-1}} = \theta' = \theta^x$ and $wxw^{-1}x^{-1} \in I(\theta) \cap U = C$. As $\overline{U} \rtimes W_1$ is a Frobenius group, $x \in C$ and $\theta' = \theta$. It follows that all the characters μ_j $(1 \leq j < p)$ are obtained in this way. If (c) does not hold, then $(p-1)^q/u = p - 1$, and so $u = (p-1)^{q-1}$, which is a contradiction because u is odd.

(d) Let $\theta_1 \in \mathrm{Irr}(\overline{H}/(H_2 \ldots H_q))$, $\theta_1 \neq 1_{\overline{H}/(H_2\ldots H_q)}$. Let $\lambda \in \mathrm{Irr}(C_U(H_1)/U')$. Let $\theta = \theta_1 \lambda$. Then θ can be identified with a character of $HC_U(H_1)$ and $I(\theta) \cap U \subset I(\theta_1) \cap U = C_U(H_1)$. Thus the character χ induced to HU from θ is an irreducible character of degree $|U : C_U(H_1)| = a$. Since $\theta_1 \neq 1_{\overline{H}/(H_2\ldots H_q)}$, $\mathrm{Ind}^M_{HU}\chi \in \mathcal{S}(H_0U')$. As each of the characters θ has a conjugates under U, we thus obtain $\dfrac{p-1}{a}|C_U(H_1) : U'| = \dfrac{p-1}{a}\dfrac{|U|}{a|U'|}$ characters χ. For such a character χ, $H_2\ldots H_q \subset \mathrm{Ker}\,\chi$, and so $H_1 \subset \mathrm{Ker}(\chi^w)$ if $w \in W_1^{\#}$. It follows that two such characters χ which are distinct are not conjugate under W_1 and so induce distinct characters of M, and that $I(\chi) \cap W_1 = 1$ for some such χ, whence $\mathrm{Ind}_{HU}^M \chi$ is irreducible. $\qquad\qquad\qquad\qquad\qquad\qquad\qquad\square$

(9.9) *Suppose that case (9.7.b) holds.*

   (a) *If $\chi \in \mathcal{X}(H_0)$, then $\chi(1)$ is divisible by u. If $\chi \in \mathcal{X}(H_0C')$, then $\chi(1) = u$ and χ is induced to HU from a linear character of HC.*

   (b) *$\mathcal{S}(H_0)$ contains exactly $p-1$ reducible characters μ_j $(1 \leq j < p)$. For $1 \leq j < p$, $\mu_j(1) = qu$ and $\mu_j \in \mathcal{S}(H_0C)$.*

   (c) *Suppose that $\mathcal{S}(H_0C')$ does not contain an irreducible character. Then $C = 1$ and $u = (p^q - 1)/(p - 1)$.*

Proof. (a) Let θ be an irreducible component of $\mathrm{Res}_H^{HU} \chi$. Then $\theta \neq 1$, and $I(\theta) \cap U = C$ since $\overline{H} \rtimes \overline{U}$ is a Frobenius group. By (1.7.a), χ is induced to HU from an irreducible character of $I(\theta) \cap HU$. Thus $\chi(1)$ is divisible by $u = |U : C|$. Suppose that $\chi \in \mathcal{X}(H_0C')$. Let $\theta \in \mathrm{Irr}\,\overline{H}$ and $\lambda \in \mathrm{Irr}\,C$ be such that $\theta\lambda$ is an irreducible component of $\mathrm{Res}_{HC}^{HU} \chi$. Then $\theta \neq 1$ and $I(\theta\lambda) \cap U = C$. Thus $\mathrm{Ind}_{HC}^{HU}(\theta\lambda)$ is irreducible, and so equal to χ. As $C' \subset \mathrm{Ker}\,\chi$, $(\theta\lambda)(1) = 1$, and so $\chi(1) = u$.

   (b) By (4.7) and Theorem (4.5), both $\mathcal{S}(H_0)$ and $\mathcal{S}(H_0C)$ contain exactly $p-1$ reducible characters μ_j $(1 \leq j < p)$. By (a), $\mu_j(1) = qu$ for $1 \leq j < p$.

   (c) Suppose that $C \neq 1$. Let $\theta \in \mathrm{Irr}\,\overline{H}$ and $\lambda \in \mathrm{Irr}\,C$ be such that $\theta \neq 1$, $\lambda \neq 1$ and $(\theta\lambda)(1) = 1$. Then, as we have seen in (a), $\chi = \mathrm{Ind}_{HC}^{HU}(\theta\lambda)$ is irreducible and $\mathrm{Ind}_{HU}^M \chi \in \mathcal{S}(H_0C')$. As $C \not\subset \mathrm{Ker}\,\chi$, it follows from (b) that $\mathrm{Ind}_{HU}^M \chi$ is irreducible, contrary to hypothesis.

   Suppose that $C = 1$. Then the Frobenius group $\overline{H} \rtimes U$ has $(p^q - 1)/u$ irreducible characters whose kernels do not contain $\overline{H}$ ([Is], Theorem 6.34). If all these characters induce reducible characters of M, then $(p^q - 1)/u = p - 1$ by (b), whence $u = (p^q - 1)/(p - 1)$. $\qquad\qquad\qquad\qquad\square$

(9.10) *Suppose that $\mathcal{S}(H_0C')$ contains no irreducible character of degree qu induced from a linear character of HC. Then case (9.7.b) holds, $\overline{H}U$ is a Frobenius group with kernel $\overline{H}$ and U is cyclic of order $(p^q - 1)/(p - 1)$. If, further, M is of Type II, then HU is a Frobenius group with kernel H.*

Proof. The first assertion follows from (9.8.c), (9.9.a, c) and (9.7). Suppose that M is of Type II. Since U is cyclic, HU is a Frobenius group with kernel H by (8.6.b II) and (8.2.b). $\qquad\qquad\qquad\qquad\qquad\qquad\qquad\qquad\qquad\square$

(9.11) *$\mathcal{S}(H_0C')$ is coherent for τ.*

Proof. By (8.15), Hypothesis (5.2) holds for $L = M$. By (9.9.a) and (5.7), $\mathcal{S}(H_0C')$ is coherent in case (9.7.b). Suppose that case (9.7.a) holds. Let $\mathcal{S}_1$

be the set of elements of $\mathcal{S}(H_0C')$ of degree qa. By (9.8.d), $\mathcal{S}_1 \neq \emptyset$ and, by (5.7), $\mathcal{S}_1$ is coherent. Let $\mathcal{S}_2$ be maximal such that $\mathcal{S}_1 \subset \mathcal{S}_2 \subset \mathcal{S}(H_0C')$, $\mathcal{S}_2$ is coherent and $\mathcal{S}_2$ is closed under complex conjugation. Let $\mathcal{S}_3 = \mathcal{S}(H_0C') - \mathcal{S}_2$. Suppose that $\mathcal{S}_3 \neq \emptyset$. Set $U_1 = C_U(H_1)$.

(9.11.1) $\mathcal{S}_2 = \mathcal{S}_1 \subset \mathcal{S}(H_0C) \cap \mathrm{Irr}(M)$, $a = (p-1)/2$, $C = U'$, *every element of $\mathcal{S}_3$ is of degree* $qu \neq qa$ *and* $|\mathcal{S}_1| = (p-1)u/a^2 = 2u/a$.

Proof. Let $\chi \in \mathcal{X}(H_0C')$. There is thus a character $\theta \in \mathrm{Irr}(HC)$ such that $H_0C' \subset \mathrm{Ker}\,\theta$ and such that χ is an irreducible component of $\mathrm{Ind}_{HC}^{HU}\,\theta$. Thus, $\chi(1) \leq |U : C|\theta(1) = u$. Let $\chi \in \mathcal{X}(H_0C')$ be such that $\mathrm{Ind}_{HU}^M \chi \in \mathcal{S}_3$. By Theorem (5.6) and by (9.8.a, d),

$$(p-1)|U : U'|q^2 \;\leq\; \sum_{\psi \in \mathcal{S}_1 \cap \mathcal{S}(H_0U') \cap \mathrm{Irr}(M)} \frac{\psi(1)^2}{\|\psi\|^2}$$

$$\leq \; \sum_{\psi \in \mathcal{S}_2} \frac{\psi(1)^2}{\|\psi\|^2}$$

$$\leq \; 2q^2 a\chi(1) \;\leq\; 2q^2 au.$$

Thus $((p-1)/2)|C : U'| \leq a$. As a divides $p-1$ and is odd, $a \leq (p-1)/2$. It follows that $C = U'$ and $a = (p-1)/2$. Furthermore, the inequalities above are equalities. Then $\mathcal{S}_2 = \mathcal{S}_1 \cap \mathcal{S}(H_0U') \cap \mathrm{Irr}(M)$. Therefore, it follows that $\mathcal{S}_2 = \mathcal{S}_1 \subset \mathcal{S}(H_0C) \cap \mathrm{Irr}(M)$, $|\mathcal{S}_1| = (p-1)|U : U'|/a^2 = (p-1)u/a^2$ and $\chi(1) = u$. By the definition of $\mathcal{S}_1$, $u \neq a$. $\square$

(9.11.2) *If $w \in W_1^\#$, then $U_1 \cap U_1^w = C$. Moreover, $u \leq a^2$.*

Proof. We may assume that $H_1^w = H_2$. Let θ be an irreducible character of $\overline{H}C$ such that $H_3 \ldots H_qC \subset \mathrm{Ker}\,\theta$ and such that the restrictions of θ to H_1 and to H_2 are $\neq 1$. Then $I(\theta) \cap U = C_U(H_1) \cap C_U(H_2) = U_1 \cap U_1^w$. Let χ be an irreducible component of $\mathrm{Ind}_{HC}^{HU}\,\theta$. By (1.6) and (1.7.c) applied to the group $(HU)/(H_0C)$, $\chi(1) = |U : I(\theta) \cap U|\theta(1) = |U : U_1 \cap U_1^w|$. By (9.11.1), $\chi(1) = u$ or a. If $\chi(1) = a$, then $U_1 \cap U_1^w = U_1$, and so W_1 normalizes U_1 and $C = \bigcap_{1 \leq i \leq q} C_U(H_i) = U_1$. Then $u = a$, which contradicts (9.11.1). Thus $\chi(1) = u$ and $U_1 \cap U_1^w = C$. The canonical mapping from $\overline{U}$ to the group $(U/U_1) \times (U/U_1^w)$ is then injective, and so $u \leq a^2$. $\square$

(9.11.3) *Let $\mathcal{S}_4$ be the set of irreducible characters in $\mathcal{S}_3 \cap \mathcal{S}(H_0C)$. Then*

$$|\mathcal{S}_4| = \frac{1}{q}\left(\frac{(p^q - 1) - (p-1)q}{u} - (p-1)\right).$$

Proof. Let n be the number of irreducible characters of $\mathcal{X}(H_0C)$ which are of degree u. By (9.11.1) and (9.8.b), $\mathcal{X}(H_0C)$ contains n characters which are of degree u and $q(p-1)u/a^2$ characters of degree a. Thus,

$$p^q u = |\overline{HU}| = |\overline{U}| + \sum_{\chi \in \mathcal{X}(H_0C)} \chi(1)^2 = u + nu^2 + q(p-1)u.$$

It follows that $n = ((p^q - 1) - (p-1)q)/u$. By (9.8.b), $p-1$ of these characters induce reducible characters of M. Each of the others has q conjugates under W_1, which proves (9.11.3). $\square$

(9.11.4) *Let* $\psi_1 \in \mathcal{S}_1$, *let* $\gamma = \mathrm{Ind}_{HU_1}^{M} 1_{HU_1}$ *and let* $\alpha = \gamma - \psi_1$. *Then* $\mathrm{Supp}(\alpha) \subset A(M)$ *and* $\|\alpha\|^2 = a + 1 + \dfrac{(q-1)a^2}{u}$.

Proof. Let $y \in U_1$. Then $C_{\overline{H}}(y) \neq 1$, and so $C_H(y) \neq 1$ by [BG], Proposition 1.5(d). Thus $HU_1 \subset A(M) \cup \{1\}$. By (4.7), $\mathrm{Supp}(\psi_1) \subset A(M) \cup \{1\}$, and so $\mathrm{Supp}(\alpha) \subset A(M)$. Since $H \subset \mathrm{Ker}\,\gamma$, γ is orthogonal to ψ_1 and $\|\alpha\|^2 = 1 + \|\gamma\|^2$. By (1.6), we may identify γ with $\mathrm{Ind}_{\overline{U_1}}^{\overline{U}W_1} 1_{\overline{U_1}}$, where $\overline{U_1}$ is the image of U_1 in $\overline{U}$. Since $\overline{U}$ is abelian,

$$(\mathrm{Ind}_{\overline{U_1}}^{\overline{U}} 1_{\overline{U_1}})(x) = \begin{cases} a & \text{if } x \in \overline{U_1}, \\ 0 & \text{if } x \in \overline{U} - \overline{U_1}. \end{cases}$$

If $x \in \overline{U}$, then $\gamma(x) = a\lambda(x)$, where $\lambda(x)$ is the number of $w \in W_1$ such that $x^w \in \overline{U_1}$. By (9.11.2),

$$\lambda(x) = \begin{cases} q & \text{if } x = 1, \\ 1 & \text{if } x \in \bigcup_{w \in W_1}(\overline{U_1}^w - \{1\}), \\ 0 & \text{otherwise.} \end{cases}$$

Thus, $\|\gamma\|^2 = \dfrac{a^2}{uq}[q^2 + q(|\overline{U_1}| - 1)] = \dfrac{a^2}{u}(q + \dfrac{u}{a} - 1) = a + \dfrac{(q-1)a^2}{u}$. $\square$

(9.11.5) *We have* $|\mathcal{S}_4| > \|\alpha\|^2$ *and* $\mathcal{S}_4 \neq \emptyset$.

Proof. Suppose that $|\mathcal{S}_4| \leq \|\alpha\|^2$. By (9.11.3) and (9.11.4),

$$\frac{p^q - 1}{qu} - \frac{p-1}{u} - \frac{p-1}{q} \leq a + 1 + \frac{(q-1)a^2}{u},$$

and so

$$\begin{aligned} p^q - 1 &\leq (p-1)q + (p-1)u + (a+1)qu + q(q-1)a^2 \\ &= (qa + 2a + q)u + q(q-1)a^2 + 2qa. \end{aligned}$$

Then $p^q - 1 \leq (q+2)a^3 + q^2a^2 + 2qa$ by (9.11.2). As $p = 2a + 1$, we obtain

$$2^q a^q + 2q(q-1)a^2 + 2aq \leq (2a+1)^q - 1 \leq (q+2)a^3 + q^2a^2 + 2qa$$

and

$$2^q a^q \leq (q+2)a^3 - (q^2 - 2q)a^2 \leq (q+2)a^3.$$

Since $q \geq 3$, it follows that $2^q \leq q+2$. But, by induction, we see that $2^x > x+2$ for all integers $x \geq 3$, a contradiction. $\square$

For $i = 1$ and $i = 3$, let τ_i be an extension of τ to $\mathbf{Z}[\mathcal{S}_i]$; that such extensions exist can be seen from (5.7).

(9.11.6) *We have that α^τ is orthogonal to $\mathcal{S}_3^{\tau_3}$.*

Proof. If λ_1, $\lambda_2 \in \mathcal{S}_3$, then $(\alpha^\tau, (\lambda_1 - \lambda_2)^\tau) = (\alpha, \lambda_1 - \lambda_2) = 0$ because $H \not\subset \operatorname{Ker} \lambda_i$. Thus, $(\alpha^\tau, \lambda_1^{\tau_3}) = (\alpha^\tau, \lambda_2^{\tau_3})$. If there is a character $\lambda \in \mathcal{S}_3$ such that $(\alpha^\tau, \lambda^{\tau_3}) \neq 0$, then all elements of $\mathcal{S}_4^{\tau_3}$ are components of α^τ, and so $\|\alpha^\tau\|^2 = \|\alpha\|^2 \geq |\mathcal{S}_4|$, which contradicts (9.11.5). $\qquad\square$

(9.11.7) *Let $\lambda_1 \in \mathcal{S}_4$ and $\beta = \lambda_1 - (u/a)\psi_1$. Then*

$$\beta^\tau = \Gamma - \frac{u}{a}\psi_1^{\tau_1} + b \sum_{\psi \in \mathcal{S}_1} \psi^{\tau_1},$$

where $\Gamma \in \mathbf{Z}[\mathcal{S}_4^{\tau_3}]$, $\|\Gamma\|^2 = 1$, and $b = 0$ or $b = 1$.

Proof. By (5.5), $\mathcal{S}_1^{\tau_1}$ is orthogonal to $\mathcal{S}_4^{\tau_3}$. Set
$$\beta^\tau = \Gamma + \textstyle\sum_{\psi \in \mathcal{S}_1} b_\psi \psi^{\tau_1} + \Delta,$$
where $\Gamma \in \mathbf{Z}[\mathcal{S}_4^{\tau_3}]$, $b_\psi \in \mathbf{Z}$ and Δ is orthogonal to $\mathcal{S}_1^{\tau_1} \cup \mathcal{S}_4^{\tau_3}$. For $\psi \in \mathcal{S}_1$, if $\psi \neq \psi_1$, then $(\beta^\tau, (\psi - \psi_1)^\tau) = (\beta, \psi - \psi_1) = u/a$, and so $b_{\psi_1} = b_\psi - u/a$. Thus,

$$\beta^\tau = \Gamma - \frac{u}{a}\psi_1^{\tau_1} + b \sum_{\psi \in \mathcal{S}_1} \psi^{\tau_1} + \Delta,$$

where $b = b_{\psi_1} + u/a$. As $(\beta^\tau, (\lambda_1 - \overline{\lambda_1})^\tau) \neq 0$, $\Gamma \neq 0$ and so $\|\Gamma\|^2 \geq 1$. But

$$\|\Gamma\|^2 + \left(\frac{u}{a} - b\right)^2 + b^2(|\mathcal{S}_1| - 1) + \|\Delta\|^2 = \|\beta\|^2 = \frac{u^2}{a^2} + 1$$

and $|\mathcal{S}_1| = 2u/a$ by (9.11.1), and so

$$\frac{u^2}{a^2} - \frac{2bu}{a} + \frac{b^2 2u}{a} + \|\Delta\|^2 \leq \frac{u^2}{a^2},$$

or $2(b^2 - b)u/a + \|\Delta\|^2 \leq 0$. Since $b \in \mathbf{Z}$, the fact that $b^2 - b \leq 0$ implies that $b = 0$ or $b = 1$, and then $\|\Delta\|^2 \leq 0$ whence $\Delta = 0$. The inequality above is thus an equality, and so $\|\Gamma\|^2 = 1$. $\qquad\square$

(9.11.8) *Conclusion.*

Since $H \subset \operatorname{Ker}\gamma$, γ is orthogonal to $\mathcal{S}$, and so $(\alpha^\tau, \beta^\tau) = (\alpha, \beta) = u/a$. As $(\alpha^\tau, (\psi - \psi_1)^\tau) = 1$ for $\psi \in \mathcal{S}_1$, $\psi \neq \psi_1$, there is an integer x for which $\alpha^\tau = x \sum_{\psi \in \mathcal{S}_1} \psi^{\tau_1} - \psi_1^{\tau_1} + \Delta$, where Δ is orthogonal to $\mathcal{S}_1^{\tau_1}$. By (9.11.6) and (9.11.7), α^τ is orthogonal to Γ and

$$(\alpha^\tau, \beta^\tau) = \left(b - \frac{u}{a}\right)(x - 1) + bx(|\mathcal{S}_1| - 1) \equiv b(x - 1) - bx = -b \pmod{\frac{u}{a}}.$$

Thus u/a divides b. But $u/a > 1$ by (9.11.1), and so $b = 0$. It follows as in (5.6.3) that $\mathcal{S}_1 \cup \{\lambda_1, \overline{\lambda_1}\}$ is coherent. But this is a contradiction to the definition of $\mathcal{S}_2$. $\qquad\square$

10. Maximal Subgroups
of Types III, IV and V

(10.1) Hypothesis. *M is a maximal subgroup of G of Type III, IV or V while M', M'', W_1, W_2 and V have the same meaning as in Definition (8.4). Denote by τ the Dade isometry relative to $(A_0(M), M, G)$ and let*

$$\mathcal{S} = \{\mathrm{Ind}_{M'}^M \theta \mid \theta \in \mathrm{Irr}\, M', \ \theta \neq 1_{M'}\}.$$

By (8.15), Hypotheses (4.6) and (5.2) hold with $L = M$ and $H = K = M'$. The symbols w_1, w_2, σ, ω_{ij}, μ_{ij}, μ_j and δ_j have the same meaning as in Hypothesis (4.6) with $L = M$.

(10.2) *Assume Hypothesis (10.1). Then there is a character $\zeta \in \mathcal{S} \cap \mathrm{Irr}\, M$ such that $\zeta(1) = w_1$.*

Proof. By (8.4.d), $W_2 \subset M''$. As M' is solvable and $M' \neq 1$, it follows that $(M'/M'') \rtimes W_1$ is a Frobenius group with kernel M'/M''. Thus the character induced to M from a non-principal character of M' of degree 1 is an irreducible character in $\mathcal{S}$ of degree w_1. $\qquad\square$

(10.3) *Assume Hypothesis (10.1). Then w_2 is prime. For $0 \leq i < w_1$ and $0 < j < w_2$, $d = \mu_{ij}(1)$ is independent of i and j, and $\delta = \delta_j$ is independent of j. Moreover, $d > 1$ and $n = (d - \delta)/w_1 \in \mathbf{N}$.*

Proof. By Theorem (8.8), there is a maximal subgroup S of G of Type II such that $|S : [S, S]| = w_2$, and so w_2 is prime. By (4.5.a), $\mu_{ij}(1)$ does not depend on i. Let j be such that $0 < j < w_2$. By (4.3.b) and (3.9.b), there is an automorphism u of $\mathbf{Q}_{|M|}$ such that $\delta_j \mu_{0j} = (\delta_1 \mu_{01})^u$. It follows that $\delta_j = \delta_1$ and $\mu_{0j}(1) = \mu_{01}(1)$. By (4.4), $d > 1$ and, by (4.3.d), $n \in \mathbf{N}$. $\qquad\square$

(10.4) Hypothesis. (a) *Assume Hypothesis (10.1) and that ζ, d, δ and n are as in (10.2) and (10.3).*

(b) *$\mathcal{S}$ is coherent and τ_1 is an extension of τ to $\mathbf{Z}[\mathcal{S}]$.*

(10.5) *Assume (10.4.a). For $0 \leq i < w_1$, $0 < j < w_2$, let $\alpha_{ij} = \mu_{ij} - \delta\mu_{i0} - n\zeta$. Then $\mathrm{Supp}(\alpha_{ij}) \subset A_0(M)$. If Hypothesis (10.4) holds, then*

$$\alpha_{ij}^\tau = \delta(\omega_{ij}^\sigma - \omega_{i0}^\sigma) - n\zeta^{\tau_1}.$$

Proof. By (4.3.c) and (4.4), $(\mu_{ij} - \delta\mu_{i0})(x) = \delta(\omega_{ij} - \omega_{i0})(x) = 0$ for $x \in W_1^{\#}$. By (4.4), $\mu_{i0}(1) = 1$, and so $\alpha_{ij}(1) = 0$ by definition of n. Thus, α_{ij} vanishes on W_1. If $z \in M - M'$, then, by (2.1), z is conjugate in M to an element of xW_2 for some $x \in W_1^{\#}$. Thus, if $z \in \mathrm{Supp}(\alpha_{ij})$, $z \in V^M$. It follows that $\mathrm{Supp}(\alpha_{ij}) \subset A_0(M)$. Assume Hypothesis (10.4). Set $(\alpha_{ij}^\tau, \zeta^{\tau_1}) = a - n$. Then

$(\alpha_{ij}^{\tau}, (\zeta - \overline{\zeta})^{\tau}) = (\alpha_{ij}, \zeta - \overline{\zeta}) = -n$ whence $(\alpha_{ij}^{\tau}, \overline{\zeta}^{\tau_1}) = a$. Let k be such that $0 < k < w_2$ and $k \neq j$. Then $\mu_k(1) = dw_1$ and

$$0 = (\alpha_{ij}, \mu_k - d\overline{\zeta}) = (\alpha_{ij}^{\tau}, \mu_k^{\tau_1} - d\overline{\zeta}^{\tau_1}) = (\alpha_{ij}^{\tau}, \mu_k^{\tau_1}) - da.$$

It follows that

$$d^2 a^2 = (\alpha_{ij}^{\tau}, \mu_k^{\tau_1})^2 \leq \|\alpha_{ij}^{\tau}\|^2 \|\mu_k^{\tau_1}\|^2 = (2 + n^2)w_1.$$

Suppose that $a \neq 0$. Then $d^2 = (w_1 n + \delta)^2 \leq w_1(2 + n^2)$, whence

$$(w_1^2 - w_1)n^2 + 2\delta w_1 n + 1 - 2w_1 \leq 0$$

and

$$n^2 - n - 1 \leq \frac{1}{2}(2n^2 + 2\delta n - 2) \leq \frac{1}{2w_1}[(w_1^2 - w_1)n^2 + 2\delta w_1 n - 2w_1] < 0.$$

Thus $n < 2$. But this is a contradiction because, by (10.3), n is an even integer and $n > 0$. Thus, $a = 0$ and $\alpha_{ij}^{\tau} = X - n\zeta^{\tau_1}$, where X is a virtual character of G orthogonal to ζ^{τ_1} such that $\|X\|^2 = 2$. By (5.3.b), (5.5) and (3.2.d), ζ^{τ_1} vanishes on V. By (3.2.c), (4.3.c) and the definition of τ, $\alpha_{ij}^{\tau} - \delta(\omega_{ij}^{\sigma} - \omega_{i0}^{\sigma})$ vanishes on V. Thus, $\psi = X - \delta(\omega_{ij}^{\sigma} - \omega_{i0}^{\sigma})$ vanishes on V. In the notation of Hypothesis (3.6), $NC(\psi) \leq 4 < 2\inf(w_1, w_2)$. By (3.8), ψ is then orthogonal to ω_{ij}^{σ} and ω_{i0}^{σ}, and so $\psi = 0$. $\square$

(10.6) *Assume Hypothesis* (10.4).

   (a) *If* $0 < j < w_2$, *then* $\mu_j^{\tau_1} = \delta \sum_{0 \leq i < w_1} \omega_{ij}^{\sigma}$. *Also,*

$$(\mu_0 - \zeta)^{\tau} = \sum_{0 \leq i < w_1} \omega_{i0}^{\sigma} - \zeta^{\tau_1}.$$

   (b) *Let* $g \in G - \tilde{A}(M)$ *have order prime to* w_1. *Then* $|\zeta^{\tau_1}(g)| \geq 1$.

Proof. (a) By (10.5),

$$1 = (\alpha_{ij}, \mu_j - d\overline{\zeta}) = (\alpha_{ij}^{\tau}, \mu_j^{\tau_1} - d\overline{\zeta}^{\tau_1}) = (\delta(\omega_{ij}^{\sigma} - \omega_{i0}^{\sigma}), \mu_j^{\tau_1})$$

for $0 < j < w_2$. By (5.8), it follows that $\mu_j^{\tau_1} = \delta \sum_{0 \leq i < w_1} \omega_{ij}^{\sigma}$. On the other hand,

$$\sum_{0 \leq i < w_1} \alpha_{ij} = \mu_j - \delta\mu_0 - nw_1\zeta = (\mu_j - d\zeta) - \delta(\mu_0 - \zeta),$$

and so

$$\begin{aligned}
\delta(\mu_0 - \zeta)^{\tau} &= (\mu_j - d\zeta)^{\tau} - \sum_{0 \leq i < w_1} \alpha_{ij}^{\tau} \\
&= \delta \sum_{0 \leq i < w_1} \omega_{ij}^{\sigma} - d\zeta^{\tau_1} - \sum_{0 \leq i < w_1} (\delta(\omega_{ij}^{\sigma} - \omega_{i0}^{\sigma}) - n\zeta^{\tau_1}) \\
&= \delta \sum_{0 \leq i < w_1} \omega_{i0}^{\sigma} - \delta\zeta^{\tau_1}.
\end{aligned}$$

(b) By definition of τ, $(\mu_0 - \zeta)^\tau$ vanishes on $G - \tilde{A}(M)$. Thus,

$$\zeta^{\tau_1}(g) = \sum_{0 \leq i < w_1} \omega_{i0}^\sigma(g).$$

By (3.9.c), $\omega_{i0}^\sigma(g) \in \mathbf{Z}$. For $i \neq 0$, $\overline{\omega_{i0}} \neq \omega_{i0}$ and, by (3.9.a), $\overline{\omega_{i0}}^\sigma = \overline{\omega_{i0}^\sigma}$. It follows that $\sum_{0 < i < w_1} \omega_{i0}^\sigma(g) \in 2\mathbf{Z}$. Then, since $\omega_{00}^\sigma = 1_G$ by (3.2.b), $\zeta^{\tau_1}(g) \in \mathbf{Z}$ and $\zeta^{\tau_1}(g) \equiv 1 \pmod{2}$. Thus $|\zeta^{\tau_1}(g)| \geq 1$.   □

(10.7) *Assume Hypothesis* (10.4). *Let S be a maximal subgroup of G of Type II. Then $[S, S]$ is a Frobenius group with kernel S_F.*

Proof. By Theorem (8.8), we may assume that $S \cap M = W$, that $S = [S, S] \rtimes W_2$ and that $C_{[S,S]}(W_2) = W_1$. Let $H = S_F$, $[S, S] = H \rtimes U$ and $\mathcal{T} = \{\mathrm{Ind}_{HU}^S \chi \mid \chi \in \mathrm{Irr}(HU), \ H \not\subset \mathrm{Ker}\,\chi\}$; again denote by τ the Dade isometry relative to $(A(S), S, G)$. By Hypothesis (10.1), M is not a Frobenius group with kernel M_F, whence, by (8.13.c4), no conjugate of S supports M. By (8.18.b), it follows that $\tilde{A}(S) \cap \tilde{A}_1(M) = \emptyset$. If $\alpha \in \mathbf{Z}[\mathcal{S}, M^\#]$ and $\beta \in \mathbf{Z}[\mathcal{T}, S^\#]$, then $\mathrm{Supp}(\alpha) \subset (M')^\# = A_1(M)$ by (8.10) and Hypothesis (10.1), and $\mathrm{Supp}(\beta) \subset A(S)$ by (8.15) and (4.7), whence $(\alpha^\tau, \beta^\tau) = 0$. Let ν_i, $0 \leq i < w_1$, be the reducible characters of S given by Theorem (4.5). Suppose that HU is not a Frobenius group with kernel H. By (9.10), (9.8.b) and (9.9.b), there is an index $r \neq 0$ such that $\nu_r \in \mathcal{T}$ and there is a character $\lambda \in \mathcal{T} \cap \mathrm{Irr}\,S$ such that $\lambda(1) = \nu_r(1)$. By (5.7), $\{\lambda, \overline{\lambda}, \nu_r, \overline{\nu_r}\}$ is coherent. Let τ_2 be an extension of τ to $\mathbf{Z}[\{\lambda, \overline{\lambda}, \nu_r, \overline{\nu_r}\}]$. Let s be such that $0 < s < w_2$, let $\alpha = \mu_s - d\zeta$ and let $\beta = \nu_r - \lambda$. By (5.8), there are indices r' and s' such that

$$\mu_s^{\tau_1} = \pm \sum_{0 \leq i < w_1} \omega_{is'}^\sigma \quad \text{and} \quad \nu_r^{\tau_2} = \pm \sum_{0 \leq j < w_2} \omega_{r'j}^\sigma.$$

Then $(\alpha^\tau, \beta^\tau) = (\pm \sum_i \omega_{is'}^\sigma - d\zeta^{\tau_1}, \pm \sum_j \omega_{r'j}^\sigma - \lambda^{\tau_2}) = 0$ and, furthermore, $((\zeta - \overline{\zeta})^\tau, (\lambda - \overline{\lambda})^\tau) = (\zeta^{\tau_1} - \overline{\zeta}^{\tau_1}, \lambda^{\tau_2} - \overline{\lambda}^{\tau_2}) = 0$. By (4.1) and (5.3.b), the characters ω_{ij}, ζ^{τ_1} and λ^{τ_2} are pairwise orthogonal. Thus, $0 = (\alpha^\tau, \beta^\tau) = \pm \|\omega_{r's'}^\sigma\|^2 = \pm 1$, which is a contradiction.   □

(10.8) Theorem. *Assume Hypothesis* (10.1). *Then $\mathcal{S}$ is not coherent.*

Proof. Suppose that $\mathcal{S}$ is coherent. We will use the notation of Hypothesis (10.4). By Theorem (8.8), there is a maximal subgroup S of G of Type II such that $S \cap M = W$, $S = [S, S] \rtimes W_2$ and $C_{[S,S]}(W_2) = W_1$. Let $H = S_F$ and $[S, S] = H \rtimes U$. Let G_0 be the set of elements $g \in G$ of order prime to w_1 such that $g \notin \tilde{A}(M)$, let $G_1 = G - (\tilde{A}(M) \cup G_0)$, let $\chi = \zeta^{\tau_1}$ and let ρ be the mapping defined in Hypothesis (7.1) with $L = M$ and $A = A(M)$. By (7.5) with $I = \{1\}$, $L_1 = M$ and $A_1 = A(M)$,

$$\frac{1}{|G|} \left(\sum_{g \in G_0 \cup G_1} |\chi(g)|^2 - |G_0 \cup G_1| \right) + \|\chi^\rho\|^2 - \frac{|A(M)|}{|M|} \leq 0.$$

Thus, by (10.6.b),

$$-\frac{|G_1|}{|G|} + \|\chi^\rho\|^2 - \frac{|A(M)|}{|M|} \le 0.$$

By (8.4.d), $(M'/M'') \rtimes W_1$ is a Frobenius group of odd order; it follows that $|M'| \ge 2w_1 + 1$. By (7.8.b), $\|\chi^\rho\|^2 \ge 1 - \dfrac{w_1}{|M'|}$. Thus,

$$\frac{w_1}{|M'|} \ge 1 - \frac{|G_1|}{|G|} - \frac{|A(M)|}{|M|} > 1 - \frac{|G_1|}{|G|} - \frac{1}{w_1}.$$

Let $x \in G_1$. Then there is an element a of prime order dividing w_1 such that $x \in C_G(a)$. By (8.11), H is a Hall subgroup of G, and so a is conjugate to an element of H. Suppose that $a \in H^\#$. By (8.6.a), $C_G(a) \subset S$ and so $x \in C_S(a)$. If $x \in HU$, then $x \in H$ by (10.7). If $x \in S - HU$, then, by (2.1), x is conjugate in S to an element of V. Thus, $G_1 \subset (H^\#)^G \cup V^G$. Since $H^\#$ is a TI-subset of G by (8.6.a),

$$|(H^\#)^G| = \frac{|G|}{|S|}(|H| - 1) \quad \text{and} \quad |V^G| = \frac{|G|}{w_1 w_2}(w_1 w_2 - w_1 - w_2 + 1).$$

Thus,

$$\frac{w_1}{|M'|} > 1 - 1 + \frac{1}{w_2} + \frac{1}{w_1} - \frac{1}{w_1 w_2} - \frac{|H|}{|S|} - \frac{1}{w_1} = \frac{1}{w_2} - \frac{1}{w_1 w_2} - \frac{1}{w_2 |U|}.$$

As UW_2 is a Frobenius group, $|U| \ge 2w_2 + 1 \ge 7$. Thus,

$$\frac{w_1 w_2}{|M'|} > 1 - \frac{1}{w_1} - \frac{1}{|U|} \ge 1 - \frac{1}{3} - \frac{1}{7} > \frac{1}{2} \quad \text{and} \quad |M'| < 2w_1 w_2.$$

But this is a contradiction because $|M'/M''| \ge 2w_1 + 1$ and $|M''| \ge w_2$. $\quad\square$

(10.9) *Assume Hypothesis* (10.1). *Let ζ be as in* (10.2). *Suppose that $w_1 < w_2$. Then $(\mu_0 - \zeta)^\tau = \sum_{0 \le i < w_1} \omega_{i0}^\sigma - \chi$, where $\chi \in \mathbf{Z}[\mathrm{Irr}\, G]$ is orthogonal to $(\mathrm{Irr}\, W)^\sigma$ and $\|\chi\|^2 = 1$.*

Proof. Set $(\mu_0 - \zeta)^\tau = \sum_{i,j} a_{ij}\omega_{ij}^\sigma - \chi$ where $a_{ij} \in \mathbf{Z}$ and $\chi \in \mathbf{Z}[\mathrm{Irr}\, G]$ is orthogonal to $(\mathrm{Irr}\, W)^\sigma$. By (2.7), $a_{00} = ((\mu_0 - \zeta)^\tau, 1_G) = (\mu_0 - \zeta, 1_M) = 1$. Since $\mu_0 - \zeta$ vanishes on V, $(\mu_0 - \zeta)^\tau$ vanishes on V by definition of τ. Now

$$\|(\mu_0 - \zeta)^\tau\|^2 = \|\mu_0 - \zeta\|^2 = w_1 + 1 < w_2.$$

By (3.8), it follows that $\sum_{i,j} a_{ij}\omega_{ij}^\sigma = \sum_{0 \le i < w_1} \omega_{i0}^\sigma$ since $a_{00} = 1$. Consequently, $\|\chi\|^2 = \|(\mu_0 - \zeta)^\tau\|^2 - \|\sum_i \omega_{i0}^\sigma\|^2 = 1$. $\quad\square$

(10.10) Theorem. *G has no maximal subgroup of Type V.*

Proof. By Theorem (10.8), it suffices to show that, if Hypothesis (10.1) holds and M is of Type V, then $\mathcal{S}$ is coherent. Set $H = M'$ and $H' = M''$. If case

(a) of Definition (8.7) holds, then $\mathcal{S}$ is coherent by Theorem (6.8). By (8.4.d) and (8.15), we see that Hypothesis (6.4) holds for the groups denoted by L, K and M in Hypothesis (6.4) being M, H and 1 respectively. By (6.5.b), we may assume that H is a non-abelian p-group for some prime number p; then $p = w_2$ because w_2 is a prime divisor of $|H|$. By (6.5.c), case (b) of Definition (8.7) does not hold. Thus, case (c) of Definition (8.7) holds, $|H| = p^3$ and w_1 divides $p + 1$.

(10.10.1) *We have $p = 2w_1 - 1$ and $w_1 < w_2$.*

Proof. Since w_1 divides $p+1$, there is an integer $k \geq 1$ such that $p = 2kw_1 - 1$. Since H is a non-abelian group of order p^3, $|H : H'| = p^2$. Then, by (6.5.a), $|H : H'| \leq 4w_1^2 + 1$. Thus, $4k^2w_1^2 - 4kw_1 + 1 \leq 4w_1^2 + 1$ and so $k^2w_1 - k \leq w_1$, whence $k^2 - 1 \leq (k^2 - 1)w_1 \leq k$, which proves that $k = 1$. $\square$

(10.10.2) $\mathcal{S} = \mathcal{S}_1 \cup \{\mu_j \mid 0 < j < p\}$, *where $\mathcal{S}_1$ consists of $(p^2 - 1)/w_1$ irreducible characters of degree w_1. In the notation of (10.3), $d = p$, $\delta = -1$ and $n = 2$.*

Proof. Since H is a non-abelian group of order p^3, $H' = Z(H)$ has order p, and $W_2 = H'$ since $W_2 \subset H'$. Since $(H/H') \rtimes W_1$ is a Frobenius group, the set $\mathcal{S}_1$ of elements of $\mathcal{S}$ which have the subgroup H' in their kernels consists of $(|H : H'| - 1)/w_1 = (p^2 - 1)/w_1$ irreducible characters of degree w_1. If $\theta \in \mathrm{Irr}\, H$, then $\theta(1)$ divides p^3 but $\theta(1)^2 \leq p^3$, whence $\theta(1) = 1$ or $\theta(1) = p$. It follows that the elements of $\mathcal{S} - \mathcal{S}_1$ have degree pw_1. Let $\mu_j = \mathrm{Ind}_H^M \theta_j$ for $0 < j < p$. Then $\sum_{0<j<p} \theta_j(1)^2 = (p - 1)p^2 = |H| - |H : H'|$. Thus $\mathcal{S} - \mathcal{S}_1 = \{\mu_j \mid 0 < j < p\}$. It follows that $d = \mu_j(1)/w_1 = p = 2w_1 - 1$, and so $\delta = -1$ and $n = 2$. $\square$

(10.10.3) *There is an extension τ_1 of τ to $\mathbf{Z}[\mathcal{S}_1]$. Let $\zeta \in \mathcal{S}_1$ and let α_{ij} be defined as in (10.5). Then $\mathrm{Supp}(\alpha_{ij}) \subset A_0(M)$ and $\alpha_{ij}^\tau = \delta(\omega_{ij}^\sigma - \omega_{i0}^\sigma) - n\zeta^{\tau_1}$.*

Proof. By (5.7), τ_1 exists. By (10.5), $\mathrm{Supp}(\alpha_{ij}) \subset A_0(M)$. If $\lambda \in \mathcal{S}_1$, $\lambda \neq \zeta$, then $(\alpha_{ij}^\tau, (\zeta - \lambda)^\tau) = (\alpha_{ij}, \zeta - \lambda) = -n$. It follows that there is an integer $a \in \mathbf{Z}$ such that $\alpha_{ij}^\tau = X - n\zeta^{\tau_1} + a\sum_{\lambda \in \mathcal{S}_1} \lambda^{\tau_1}$, where X is a virtual character of G orthogonal to $\mathcal{S}_1^{\tau_1}$. But then

$$(a - n)^2 + (|\mathcal{S}_1| - 1)a^2 = \left\| -n\zeta^{\tau_1} + a\sum_{\lambda \in \mathcal{S}_1} \lambda^{\tau_1} \right\|^2 \leq \|\alpha_{ij}^\tau\|^2 = 2 + n^2,$$

and so $|\mathcal{S}_1|a^2 - 2an - 2 \leq 0$. As $|\mathcal{S}_1| = (p^2 - 1)/w_1 = 4(w_1 - 1) \geq 8$ and $n = 2$, $8a^2 - 4a - 2 \leq 0$. Since $a \in \mathbf{Z}$, it follows that $a = 0$. We then have $\|X\|^2 = \|\alpha_{ij}^\tau\|^2 - n^2 = 2$, and we conclude as in (10.5). $\square$

(10.10.4) $\mathcal{S}$ *is coherent.*

Proof. Let $0 < j < w_2$ and let $\zeta \in \mathcal{S}_1$. As $(\alpha_{ij}, \mu_0 - \zeta) = -\delta + n$, it follows from (10.9), (10.10.1) and (10.10.3) that $(\mu_0 - \zeta)^\tau = \sum_{0 \leq i < w_1} \omega_{i0}^\sigma - \zeta^{\tau_1}$. Since

$$n = \frac{d - \delta}{w_1} \text{ by (10.3)},$$

$$\sum_{0 \leq i < w_1} \alpha_{ij} = \sum_i \mu_{ij} - \delta \sum_i \mu_{i0} - n w_1 \zeta = \mu_j - \delta \mu_0 - (d - \delta)\zeta$$

and so $(\mu_j - d\zeta)^\tau = \delta(\mu_0 - \zeta)^\tau + \sum_{0 \leq i < w_1} \alpha_{ij}^\tau$. Then, by (10.10.3),

$$(\mu_j - d\zeta)^\tau = \delta \sum_i \omega_{i0}^\sigma - \delta\zeta^{\tau_1} + \delta \sum_i (\omega_{ij}^\sigma - \omega_{i0}^\sigma) - n w_1 \zeta^{\tau_1} = \delta \sum_i \omega_{ij}^\sigma - d\zeta^{\tau_1}.$$

It follows that $\mathcal{S}$ is coherent. $\qquad\qquad\qquad\qquad\qquad\qquad\qquad\qquad\square$

(10.11) *Suppose that case* (b) *of Theorem* (8.8) *holds. Then* $|W_1|$ *and* $|W_2|$ *are prime. If* M *is a maximal subgroup of* G *of Type II, then, in the notation of Hypotheses* (9.2) *and* (9.5), *H is an elementary abelian group of order* p^q, *where* $p = |W_2|$, *and the set* $\mathcal{S}$ *of Hypothesis* (9.5) *is coherent.*

Proof. The first assertion follows from (8.6.a) and Theorem (10.10). Suppose that M is of Type II. By (9.3), $|H| = p^q$, where $p = |W_2|$. As p is prime, (9.4) and (9.6) then show that $H_0 = 1$ and that H is elementary abelian. Since M is of Type II, U is abelian and so $C' = 1$ and $\mathcal{S}(H_0 C') = \mathcal{S}$. Thus, by (9.11), $\mathcal{S}$ is coherent. $\qquad\qquad\qquad\qquad\qquad\qquad\qquad\qquad\square$

11. Maximal Subgroups
of Types III and IV

(11.1) *Let p and q be odd prime numbers, $p \neq q$. Then $p^q > 4q^2 + 1$.*

Proof. Suppose that $q = 3$. Then $p^q \geq 5^3 > 37 = 4q^2 + 1$. Suppose that $q \geq 5$. Since $p \geq 3$, it suffices to show that $3^x > 4x^2 + 1$ for any integer $x \geq 5$. First of all, $3^5 = 243 > 101 = 4(5^2) + 1$. Suppose that $x \geq 5$ and that $3^x > 4x^2 + 1$. Then $3^{x+1} > 3(4x^2 + 1)$. But $3(4x^2 + 1) - 4(x+1)^2 - 1 = 8x^2 - 8x - 2 \geq 0$, and so $3^{x+1} > 4(x+1)^2 + 1$. $\qquad\qquad\square$

(11.2) Hypothesis. *Assume Hypothesis* (10.1) *with M of Type III or IV. We assume that H and U have the same meaning as in Definition* (8.4) *and that $H' = [H, H]$, $U' = [U, U]$ and $C = C_U(H)$. Let H_0 be a normal subgroup of M contained in H which satisfies* (9.4). *Set $p = |W_2|$ and $q = |W_1|$. For $X \subset M$, $\mathcal{S}(X) = \{\chi \in \mathcal{S} \mid X \subset \operatorname{Ker} \chi\}$.*

We remark that the set which is denoted by $\mathcal{S}$ in Hypothesis (9.5) is denoted by $\mathcal{S} - \mathcal{S}(H)$ here.

(11.3) *Assume Hypothesis* (11.2). *Then $\mathcal{S}(H_0 C)$ is not coherent.*

Proof. Suppose that $\mathcal{S}(H_0 C)$ is coherent. We show that the hypotheses of Theorem (6.3) hold with the symbols L, K, M, H and H_1 of Theorem (6.3) replaced by M, M', 1, HC and $H_0 C$ respectively. Statement (6.3.a) holds because HC is nilpotent, (6.3.b) comes from the hypothesis on $\mathcal{S}(H_0 C)$ and (6.3.c) follows from (9.6) and (11.1). By Theorem (6.3), $\mathcal{S}$ is coherent, which contradicts Theorem (10.8). $\qquad\qquad\square$

(11.4) *Assume Hypothesis* (11.2). *Let H_1 be a normal subgroup of M, strictly contained in M', such that $\mathcal{S}(H_1)$ is coherent. Then $2q|U/C| \geq |M'/H_1| - 1$.*

Proof. We see that the hypothesis of (6.2) holds with the symbols L, K, A, B, C and D of (6.2) replaced by M, M', H_1, $H_0 C$, HC and HC respectively. By (6.2), $2q|U/C| = 2|M : HC| \geq |M' : H_1| - 1$. $\qquad\qquad\square$

(11.5) *Assume Hypothesis* (11.2). *Then $M'' = HC$.*

Proof. By (8.4.c) and (8.5.a), $M'' \subset HC$. Since M'/M'' is abelian, $\mathcal{S}(M'')$ is coherent by (5.7). By (11.4),

$$|M'/M''| = |HU : HC||HC : M''| \leq 2q|U/C| + 1.$$

Thus, $|HC : M''| \leq 2q + |C|/|U| < 2q + 1$. But W_1 acts fixed-point-freely on $(HC)/M''$ by (8.4.d), and so $|HC : M''| = 1$. $\qquad\qquad\square$

(11.6) *Assume Hypothesis* (11.2). *Then H is a p-group, U centralizes H_0, $H_0 = H'$ and $C = U'$.*

Proof. By (9.3), U centralizes $O_{p'}(H)$. If $O_{p'}(H) \neq 1$, it follows that M' has a quotient of prime order dividing $|H|$. But $|M' : M''|$ is prime to $|H|$ by (11.5). Thus H is a p-group. By (9.6), $C_{H_0}(W_1) = 1$, and so U centralizes H_0 by (9.1). By [BG], Proposition 1.6(d), $H/H' = C_{H/H'}(U) \times [H/H', U]$. If $C_{H/H'}(U) \neq 1$, it follows that M'/H' has a quotient of order p, which contradicts (11.5). But $H' \subset H_0$ and U centralizes H_0/H', whence $H_0 = H'$. By (11.5), $C \subset M'' \cap U \subset (HU') \cap U = U'$, and so, by (8.5.b), $C = U'$. $\qquad\square$

(11.7) *Assume Hypothesis* (11.2). *Then H is an elementary abelian p-group of order p^q and $H_0 = 1$.*

Proof. By (9.6), it suffices to show that $H_0 = 1$. Suppose that $H_0 \neq 1$. Since H is nilpotent, $[H, H_0] \subsetneq H_0$. By [BG], Lemma 1.22, there is a normal subgroup Q of H such that $[H, H_0] \subset Q \subset H_0$ and $|H_0 : Q| = p$. By the choice of Q, $H_0/Q \subset Z(H/Q)$. Let $\overline{H} = H/H_0$; we interpret $\overline{H}$ as a vector space over $\mathbf{F}_p$. The commutator identities $[a, bc] = [a, c][a, b]^c$ and $[ab, c] = [a, c]^b[b, c]$, which are easily checked, show that the mapping $(x, y) \mapsto [x, y]$ induces an alternating $\mathbf{F}_p$-bilinear mapping from $\overline{H} \times \overline{H}$ to H_0/Q, which we also denote by $(x, y) \mapsto [x, y]$. Since U acts on H and centralizes H_0, we also have $[x^u, y^u] = [x, y]^u = [x, y]$ for x, $y \in \overline{H}$ and $u \in U$.

Suppose that U acts irreducibly on $\overline{H}$. The set N of $x \in \overline{H}$ such that $[x, y] = 0$ for all $y \in \overline{H}$ is a subspace of $\overline{H}$ normalized by U, and so $N = 0$ or $N = \overline{H}$. If $N = 0$, then, by [L], Chapter XIV, Theorem 9.1, $q = \dim \overline{H}$ is even, which is a contradiction. Thus $N = \overline{H}$, and $[x, y] = 0$ for all x, $y \in \overline{H}$. But this is a contradiction because $H_0 = H'$ is generated by the elements $[x, y]$ for x, $y \in H$.

Suppose that U does not act irreducibly on $\overline{H}$. Then case (a) of (9.7) holds. Let $\overline{H} = H_1 \oplus \cdots \oplus H_q$, where $|H_i| = p$ and H_i is normalized by U, and let φ_i be the homomorphisms from U to $(\mathbf{Z}/p\mathbf{Z})^*$ for which $h^u = \varphi_i(u)h$ for $h \in H_i$ and $u \in U$. Let $1 \leq i, j \leq q$. Suppose that $\varphi_i \varphi_j = 1$. As we have seen in the proof of (9.7), there is an element $w \in W_1$ such that $\varphi_j(u) = \varphi_i(wuw^{-1})$ for all $u \in U$. Thus, for all $u \in U$, $\varphi_i(u)\varphi_i(wuw^{-1}) = \varphi_i(wuw^{-1})\varphi_i(w^2uw^{-2}) = 1$, and so $\varphi_i(u) = \varphi_i(w^2uw^{-2})$. As w is a power of w^2, it follows that, for all $u \in U$, $\varphi_i(u) = \varphi_i(wuw^{-1}) = \varphi_j(u)$, and so $\varphi_i(u)^2 = 1$. As $|U|$ is odd, we then have $\varphi_i(u) = 1$ for all $u \in U$. But then $\varphi_k = 1$ for all k and U centralizes $\overline{H}$, which contradicts (9.4.b). Thus $\varphi_i \varphi_j \neq 1$. Let $u \in U$ be such that $\varphi_i(u)\varphi_j(u) \neq 1$. If $x \in H_i$ and $y \in H_j$, then $[x, y] = [x^u, y^u] = \varphi_i(u)\varphi_j(u)[x, y]$, and so $[x, y] = 0$. This proves that $[x, y] = 0$ for all x, $y \in \overline{H}$, which is a contradiction because H' is generated by the elements $[x, y]$ for x, $y \in H$. $\qquad\square$

(11.8) *Assume Hypothesis* (11.2). *Let $\zeta \in \mathcal{S}(HC)$. Then $(\mu_0 - \zeta)^\tau - \sum_{0 \leq i < q} \omega_{i0}^\sigma$ is not orthogonal to* $(\mathrm{Irr}\, W)^\sigma$.

Proof. Let $\mathcal{S}_1 = \mathcal{S}(HC)$ and set $u = |U/C|$. Since $(U/C) \rtimes W_1$ is a Frobenius group with abelian kernel U/C, $\mathcal{S}_1$ consists of $(u-1)/q$ irreducible characters of degree q. Let d, δ and n be as in (10.3) and let $\alpha_{ij} = \mu_{ij} - \delta\mu_{i0} - n\zeta$ for

$0 \leq i < q$, $0 < j < p$. Let τ_1 be an extension of τ to $\mathbf{Z}[\mathcal{S}_1]$, which exists by (5.7).

(11.8.1) *We have* $d = u$, $\delta = 1$ *and* $n = |\mathcal{S}_1| = \dfrac{u-1}{q}$.

Proof. By (9.8) and (9.9), $\mu_j(1) = qu$ for $j \neq 0$, and so $d = \mu_{ij}(1) = u$. Since $(U/C) \rtimes W_1$ is a Frobenius group, $u \equiv 1 \pmod{q}$ and so $\delta = 1$ and $n = (d - \delta)/q = (u - 1)/q$. $\qquad\square$

(11.8.2) *Let* $0 \leq i < q$, $0 < j < p$. *Then* $\alpha_{ij}^{\tau} = X - n\zeta^{\tau_1} + a \sum_{\lambda \in \mathcal{S}_1} \lambda^{\tau_1}$, *where* $X \in \mathbf{Z}[\mathrm{Irr}\, G]$ *is orthogonal to* $\mathcal{S}_1^{\tau_1}$ *and* $a = 0$, 1 *or* 2. *If* $a = 0$ *or* 2, *then* $X = \omega_{ij}^{\sigma} - \omega_{i0}^{\sigma}$.

Proof. By (10.5), α_{ij}^{τ} is defined. If $\lambda \in \mathcal{S}_1$, $\lambda \neq \zeta$, then $(\alpha_{ij}^{\tau}, (\zeta - \lambda)^{\tau}) = (\alpha_{ij}, \zeta - \lambda) = -n$. There is thus an integer $a \in \mathbf{Z}$ and a virtual character $X \in \mathbf{Z}[\mathrm{Irr}\, G]$ orthogonal to $\mathcal{S}_1^{\tau_1}$ such that $\alpha_{ij}^{\tau} = X - n\zeta^{\tau_1} + a \sum_{\lambda \in \mathcal{S}_1} \lambda^{\tau_1}$. Then

$$(a - n)^2 + (|\mathcal{S}_1| - 1)a^2 = \| -n\zeta^{\tau_1} + a \sum_{\lambda \in \mathcal{S}_1} \lambda^{\tau_1} \|^2 \leq \|\alpha_{ij}^{\tau}\|^2 = n^2 + 2.$$

Thus, $|\mathcal{S}_1|a^2 - 2an \leq 2$. By (11.8.1), $n(a^2 - 2a) \leq 2$, whence $0 \leq a \leq 2$. If $a = 0$ or 2, then $\| -n\zeta^{\tau_1} + a \sum_{\lambda \in \mathcal{S}_1} \lambda^{\tau_1} \|^2 = n^2$ and so $\|X\|^2 = 2$. We then see as in (10.5) that $X = \omega_{ij}^{\sigma} - \omega_{i0}^{\sigma}$. $\qquad\square$

(11.8.3) *We have that* $\beta = \alpha_{ij}^{\tau} - (\omega_{ij}^{\sigma} - \omega_{i0}^{\sigma}) + n\zeta^{\tau_1}$ *is independent of* i *and* j *for* $j \neq 0$, *and* β *is real.*

Proof. Let $\beta_{ij} = \alpha_{ij}^{\tau} - (\omega_{ij}^{\sigma} - \omega_{i0}^{\sigma}) + n\zeta^{\tau_1}$. Let k be such that $0 < k < p$. Then $(\alpha_{ij} - \alpha_{ik})^{\tau} = (\mu_{ij} - \mu_{ik})^{\tau} = \omega_{ij}^{\sigma} - \omega_{ik}^{\sigma}$ by (4.8), and so β_{ij} does not depend on j. On the other hand,

$$(\alpha_{ij} - \alpha_{0j})^{\tau} = (\mu_{ij} - \mu_{i0} - \mu_{0j} + \mu_{00})^{\tau} = \omega_{ij}^{\sigma} - \omega_{i0}^{\sigma} - \omega_{0j}^{\sigma} + \omega_{00}^{\sigma}$$

by (4.10), and so $\beta_{ij} = \beta_{0j}$ and β_{ij} is independent of i and j. Let j, $k \neq 0$ be such that $\overline{\omega_{0j}} = \omega_{0k}$. Then, by (3.9.a), (4.3.b) and (5.9),

$$
\begin{aligned}
\overline{\beta_{0j}} &= \overline{\alpha_{0j}}^{\tau} - (\overline{\omega_{0j}}^{\sigma} - \omega_{00}^{\sigma}) + n\overline{\zeta}^{\tau_1} \\
&= (\mu_{0k} - \mu_{00} - n\overline{\zeta})^{\tau} - (\omega_{0k}^{\sigma} - \omega_{00}^{\sigma}) + n\overline{\zeta}^{\tau_1} \\
&= (\mu_{0k} - \mu_{00} - n\zeta)^{\tau} + n(\zeta - \overline{\zeta})^{\tau} - (\omega_{0k}^{\sigma} - \omega_{00}^{\sigma}) + n\overline{\zeta}^{\tau_1} \\
&= \beta_{0k}.
\end{aligned}
$$

Thus β is real. $\qquad\square$

Now assume that $(\mu_0 - \zeta)^{\tau} - \sum_{0 \leq i < q} \omega_{i0}^{\sigma}$ is orthogonal to $(\mathrm{Irr}\, W)^{\sigma}$.

(11.8.4) *We may assume that* $(\mu_0 - \zeta)^{\tau} = \sum_{0 \leq i < q} \omega_{i0}^{\sigma} - \zeta^{\tau_1}$.

Proof. Set $(\mu_0 - \zeta)^{\tau} = \sum_i \omega_{i0}^{\sigma} - \chi$. Then $\|\chi\|^2 = \|\mu_0 - \zeta\|^2 - q = 1$. Considering $((\mu_0 - \zeta)^{\tau}, (\overline{\zeta} - \zeta)^{\tau})$, we see that $\chi = \zeta^{\tau_1}$ or $-\overline{\zeta}^{\tau_1}$. If $|\mathcal{S}_1| > 2$ and $\lambda \in \mathcal{S}_1 - \{\zeta, \overline{\zeta}\}$, then $((\mu_0 - \zeta)^\tau, (\lambda - \zeta)^\tau) = 1$, and so $\chi = \zeta^{\tau_1}$. If $|\mathcal{S}_1| = 2$ and $\chi = -\overline{\zeta}^{\tau_1}$, then we may replace ζ^{τ_1} and $\overline{\zeta}^{\tau_1}$ by $-\overline{\zeta}^{\tau_1}$ and $-\zeta^{\tau_1}$ respectively. $\qquad\square$

(11.8.5) $a = 0$.

Proof. Note that

$$
\begin{aligned}
((\mu_0 - \zeta)^\tau, \alpha_{ij}^\tau) &= \left(\sum_r \omega_{r0}^\sigma - \zeta^{\tau_1}, \beta + \omega_{ij}^\sigma - \omega_{i0}^\sigma - n\zeta^{\tau_1}\right) \\
&= \left(\sum_r \omega_{r0}^\sigma, \beta\right) - 1 - a + n
\end{aligned}
$$

and

$$
\begin{aligned}
(\mu_0 - \zeta, \alpha_{ij}) &= \left(\sum_r \mu_{r0} - \zeta, \mu_{ij} - \mu_{i0} - n\zeta\right) \\
&= -1 + n,
\end{aligned}
$$

whence $a = (\sum_r \omega_{r0}^\sigma, \beta)$. But $(\beta, 1_G) = (\alpha_{ij}^\tau, 1_G) = (\alpha_{ij}, 1_M) = 0$ for $i \neq 0$. As β is real, it follows that a is even. By (11.8.2), $X = \omega_{ij}^\sigma - \omega_{i0}^\sigma$. Thus, $\beta = a \sum_{\lambda \in \mathcal{S}_1} \lambda^{\tau_1}$ and, by (5.3.b), $a = (\sum_r \omega_{r0}^\sigma, \beta) = 0$. $\qquad\square$

(11.8.6) *Conclusion of the proof of* (11.8).

Let j be such that $0 < j < p$. By (11.8.2) and (11.8.5), $\alpha_{ij}^\tau = \omega_{ij}^\sigma - \omega_{i0}^\sigma - n\zeta^{\tau_1}$ for all i. It follows that

$$
\begin{aligned}
(\mu_j - d\zeta)^\tau &= (\mu_0 - \zeta)^\tau + \sum_i \alpha_{ij}^\tau = \sum_i \omega_{i0}^\sigma - \zeta^{\tau_1} + \sum_i (\omega_{ij}^\sigma - \omega_{i0}^\sigma) - nq\zeta^{\tau_1} \\
&= \sum_i \omega_{ij}^\sigma - d\zeta^{\tau_1}.
\end{aligned}
$$

Let $\mathcal{S}_2 = \mathcal{S}(C) - \mathcal{S}(HC)$. By (9.11), $\mathcal{S}(H_0C') - \mathcal{S}(HC')$ is coherent, whence $\mathcal{S}_2$ is coherent by (11.7), and $\mu_k \in \mathcal{S}_2$ for $k \neq 0$ by (9.8.b) and (9.9.b). Let τ_2 be an extension of τ to $\mathbf{Z}[\mathcal{S}_2]$. By (5.3.b) and (5.5), $\mathcal{S}_1^{\tau_1}$ is orthogonal to $\mathcal{S}_2^{\tau_2}$. If $(\mu_j - d\zeta)^\tau = \mu_j^{\tau_2} - d\zeta^{\tau_1}$, it follows that $\mathcal{S}(C) = \mathcal{S}_1 \cup \mathcal{S}_2$ is coherent, which contradicts (11.3). Thus it suffices to show that $\mu_j^{\tau_2} = \sum_i \omega_{ij}^\sigma$. If $\mathcal{S}_2 \cap \mathrm{Irr}\, M = \emptyset$, then, by Theorem (4.9), we may define τ_2 by $\mu_j^{\tau_2} = \sum_i \omega_{ij}^\sigma$. If $\lambda \in \mathcal{S}_2 \cap \mathrm{Irr}\, M$, then

$$
((\lambda(1)\mu_j - \mu_j(1)\lambda)^\tau, (\mu_j - d\zeta)^\tau) = (\lambda(1)\mu_j^{\tau_2} - \mu_j(1)\lambda^{\tau_2}, \sum_i \omega_{ij}^\sigma - d\zeta^{\tau_1}) \neq 0,
$$

and so $\mu_j^{\tau_2} = \sum_i \omega_{ij}^\sigma$ by (5.8). $\qquad\square$

(11.9) *Assume Hypothesis* (11.2). *Let* $\zeta \in \mathcal{S}(HC)$. *Then:*

    (a) $(\mu_0 - \zeta)^\tau - \sum_{0 \leq j < p} \omega_{0j}^\sigma$ *is orthogonal to* $(\mathrm{Irr}\, W)^\sigma$.

    (b) $q > p$.

    (c) *Case* (b) *of* (9.7) *holds and* M *is of Type III.*

Proof. (a) Set $(\mu_0 - \zeta)^\tau = \sum_{i,j} a_{ij}\omega_{ij}^\sigma + \chi$, where χ is orthogonal to $(\mathrm{Irr}\, W)^\sigma$. By (2.7), $a_{00} = ((\mu_0 - \zeta)^\tau, 1_G) = (\mu_0 - \zeta, 1_M) = 1$. Let α be an automorphism of $\mathbf{Q}_{|G|}$. By its definition, τ commutes with α. Thus,

$$(\mu_0 - \zeta)^{\tau\alpha} = (\mu_0 - \zeta)^\tau + (\zeta - \zeta^\alpha)^\tau$$

and, by (5.3.b), we then have $\sum a_{ij}\omega_{ij}^{\sigma\alpha} = \sum a_{ij}\omega_{ij}^\sigma$. Since p and q are prime, it then follows from (3.9.b) that $a_{i0} = a_{10}$ and $a_{0j} = a_{01}$ for $i \neq 0$ and $j \neq 0$. Moreover, $\mu_0 - \zeta$ vanishes on V, and so, by (3.7), $a_{ij} = a_{i0} + a_{0j} - a_{00}$ for all i, j. Then

$$\left\| \sum a_{ij}\omega_{ij}^\sigma \right\|^2 = 1 + (q-1)a_{10}^2 + (p-1)a_{01}^2 + (p-1)(q-1)a_{11}^2.$$

Also $\chi \neq 0$ since $((\mu_0 - \zeta)^\tau, (\zeta - \overline{\zeta})^\tau) \neq 0$, and so

$$\left\| \sum a_{ij}\omega_{ij}^\sigma \right\|^2 \leq \|\mu_0 - \zeta\|^2 - 1 = q.$$

If $a_{11} \neq 0$, then $\| \sum a_{ij}\omega_{ij}^\sigma \|^2 \geq (p-1)(q-1) \geq 2(q-1)$ and so $2q - 2 \leq q$, which is a contradiction. If $a_{10} \neq 0$, then

$$q + (p-1)a_{01}^2 \leq 1 + (q-1)a_{10}^2 + (p-1)a_{01}^2 \leq q,$$

and so $a_{01} = 0$ and $a_{10} - 1 = a_{10} + a_{01} - a_{00} = a_{11} = 0$. In this case, $\sum a_{ij}\omega_{ij}^\sigma = \sum_i \omega_{i0}^\sigma$, which contradicts (11.8). Thus, $a_{10} = 0$ and $a_{10} + a_{01} - a_{00} = a_{11} = 0$, whence $a_{01} = 1$.

(b) This follows from (10.9) and (11.8).

(c) Suppose that case (a) of (9.7) holds; let a be as in (9.7.a). By (11.6) and (11.7), $U' = C$ and $H_0 = 1$. By (9.8.d), there is an irreducible character $\lambda \in \mathcal{S}(C) - \mathcal{S}(HC)$ such that $\lambda(1) = qa$. Let j be such that $0 < j < p$. By (9.8.b), $\mu_j \in \mathcal{S}(C) - \mathcal{S}(HC)$ and $\mu_j(1) = qu$. By (9.11), $\mathcal{S}(C) - \mathcal{S}(HC)$ is coherent. Let τ_2 be an extension of τ to $\mathbf{Z}[\mathcal{S}(C) - \mathcal{S}(HC)]$. By (5.8), there is an index $k \neq 0$ such that

$$\left(\mu_j - \frac{u}{a}\lambda\right)^\tau = \pm \sum_i \omega_{ik}^\sigma - \frac{u}{a}\lambda^{\tau_2}.$$

If χ is as in the proof of (a), then

$$0 = \left(\mu_0 - \zeta, \mu_j - \frac{u}{a}\lambda\right) = \left((\mu_0 - \zeta)^\tau, \left(\mu_j - \frac{u}{a}\lambda\right)^\tau\right) = \pm 1 - \frac{u}{a}(\chi, \lambda^{\tau_2}).$$

Since $\dfrac{u}{a}$ and (χ, λ^{τ_2}) are integers, $a = u$. By (9.7.a), a divides $p - 1$, and $q < u = a < p$ since W_1 acts fixed-point-freely on U/C. This contradicts (b) and so case (b) of (9.7) holds. It follows that U/C is cyclic. As $C = U'$ and U is nilpotent, U is then cyclic, whence M is of Type III. $\qquad\square$

12. Maximal Subgroups of Type I

(12.1) Hypothesis. *Let L be a maximal subgroup of G of Type I, $H = L_F$ and $H' = [H, H]$. Let $\mathcal{S} = \{\operatorname{Ind}_H^L \theta \mid \theta \in \operatorname{Irr} H, \ \theta \neq 1_H\}$ and let τ be the Dade isometry relative to $(A(L), L, G)$.*

(12.2) *Assume Hypothesis* (12.1).

(a) *Let $\chi \in \mathcal{S}$. There is a subset $S(\chi)$ of $\operatorname{Irr} L$ such that $\chi = \sum_{\varphi \in S(\chi)} \varphi$, and $\varphi(1)$ is independent of φ for $\varphi \in S(\chi)$. Moreover, τ is defined on $\mathbf{Z}[\bigcup_{\chi \in \mathcal{S}} S(\chi), L^{\#}]$.*

(b) *Hypothesis* (5.2) *holds with the isometry τ of Hypothesis* (5.2) *being the restriction of τ to $\mathbf{Z}[\mathcal{S}, L^{\#}]$. If $\chi \in \mathcal{S}$ and $\varphi \in S(\chi)$, $(\varphi - \overline{\varphi})^\tau = \sum_{\alpha \in R_1(\varphi)} \alpha$, where $R_1(\varphi)$ is an orthonormal subset of $\mathbf{Z}[\operatorname{Irr} G]$ of cardinality 2, and $R(\chi) = \bigcup_{\varphi \in S(\chi)} R_1(\varphi)$.*

Proof. (a) The first assertion follows from (8.2.c) and (1.7.c). If $\varphi \in S(\chi)$, then $(\operatorname{Res}_H^L \varphi, 1_H) = 0$ by (1.5.a), and so $H \not\subseteq \operatorname{Ker} \varphi$. By (1.2), it follows that $\operatorname{Supp}(\varphi) \subset A(L) \cup \{1\}$, and so τ is defined on $\mathbf{Z}[\bigcup_{\chi \in \mathcal{S}} S(\chi), L^{\#}]$.

(b) By (a), τ is defined on $\mathbf{Z}[\mathcal{S}, L^{\#}]$. If $\chi \in \mathcal{S}$, then $\overline{\chi} \in \mathcal{S}$ and, by (1.5.e), $\overline{\chi} \neq \chi$. The elements of $\mathcal{S}$ are pairwise orthogonal by (1.5.c). Let $\chi \in \mathcal{S}$. By (1.4) applied to $S(\chi) \cup S(\overline{\chi})$, there are pairwise distinct irreducible characters μ_φ and μ'_φ of G and an integer $\varepsilon = \pm 1$ such that, for $\varphi \in S(\chi)$, $(\varphi - \overline{\varphi})^\tau = \varepsilon(\mu_\varphi - \mu'_\varphi)$. Then $(\chi - \overline{\chi})^\tau = \sum_{\varphi \in S(\chi)} \varepsilon(\mu_\varphi - \mu'_\varphi)$ and (5.2.d) holds with $R(\chi) = \{\varepsilon \mu_\varphi, -\varepsilon \mu'_\varphi \mid \varphi \in S(\chi)\}$. Let $\chi_1 \in \mathcal{S}$ be such that χ_1 is orthogonal to $\{\chi, \overline{\chi}\}$. If $\varphi \in S(\chi)$ and $\varphi_1 \in S(\chi_1)$, then $((\varphi - \overline{\varphi})^\tau, (\varphi_1 - \overline{\varphi_1})^\tau) = (\varphi - \overline{\varphi}, \varphi_1 - \overline{\varphi_1}) = 0$. By (4.1), $R(\chi)$ is thus orthogonal to $R(\chi_1)$ and so (5.2.e) holds. $\square$

(12.3) *Let L_1 and L_2 be non-conjugate maximal subgroups of G of Type I. For $i = 1, 2$, assume that Hypothesis* (12.1) *holds with $(L_i, H_i, \mathcal{S}_i, \tau_i)$ in place of $(L, H, \mathcal{S}, \tau)$. If $\chi_1 \in \mathcal{S}_1$ and $\chi_2 \in \mathcal{S}_2$, then $R(\chi_1)$ and $R(\chi_2)$ are orthogonal.*

Proof. By (8.18.c), we may assume that $\widetilde{A}(L_1) \cap \widetilde{A_1}(L_2) = \emptyset$. Let $\alpha \in R(\chi_1)$. It suffices to show that $(\alpha, (\chi_2 - \overline{\chi_2})^{\tau_2}) = 0$. By (12.2) and (5.9), there is a character $\varphi \in S(\chi_1)$ such that $(\varphi - \overline{\varphi})^{\tau_1} = \alpha - \overline{\alpha}$. But

$$((\varphi - \overline{\varphi})^{\tau_1}, (\chi_2 - \overline{\chi_2})^{\tau_2}) = 0$$

since $\operatorname{Supp}((\varphi - \overline{\varphi})^{\tau_1}) \subset \widetilde{A}(L_1)$ and $\operatorname{Supp}((\chi_2 - \overline{\chi_2})^{\tau_2}) \subset \widetilde{A_1}(L_2)$. Thus,

$$(\alpha, (\chi_2 - \overline{\chi_2})^{\tau_2}) = (\overline{\alpha}, (\chi_2 - \overline{\chi_2})^{\tau_2}) = (\alpha, (\overline{\chi_2} - \chi_2)^{\tau_2}) = -(\alpha, (\chi_2 - \overline{\chi_2})^{\tau_2}),$$

whence $(\alpha, (\chi_2 - \overline{\chi_2})^{\tau_2}) = 0$. $\square$

(12.4) *Assume Hypothesis* (12.1). *Let $\psi \in \operatorname{CF}(G)$ be such that ψ is orthogonal to $R(\chi)$ for all $\chi \in \mathcal{S}$. If $x \in L - H$, then ψ is constant on xH.*

Proof. Let $\theta \in \operatorname{Irr} H$ be such that $\theta \neq 1_H$, $\chi = \operatorname{Ind}_H^L \theta$ and $\varphi_1, \varphi_2 \in S(\chi)$. By (12.2.a), $(\varphi_i, \chi) = (\operatorname{Res}_H^L \varphi_i, \theta) = 1$. By [Is], Theorem 6.2, it follows that $\operatorname{Res}_H^L \varphi_i$ is the sum of the conjugates of θ in L. We can then conclude that $\operatorname{Supp}(\varphi_1 - \varphi_2) \subset A(L) - H^{\#}$. By (8.12.c), $A(L) - H^{\#}$ is a TI-subset of G. If $\varphi_1 \neq \varphi_2$, this set is non-empty, and, since G is simple and L is maximal in G, its normalizer is L. Then $(\varphi_1 - \varphi_2)^{\tau} = \operatorname{Ind}_L^G(\varphi_1 - \varphi_2)$ by [Is], Lemma 7.7, and the definition of τ. By (1.4), $\{\varphi_1, \varphi_2, \overline{\varphi_1}, \overline{\varphi_2}\}$ is coherent, and so $(\varphi_1 - \varphi_2)^{\tau} \in \mathbf{Z}[R(\chi)]$. Then $(\operatorname{Res}_L^G \psi, \varphi_1 - \varphi_2) = (\psi, (\varphi_1 - \varphi_2)^{\tau}) = 0$. On the other hand, if $\varphi \in \operatorname{Irr} L$ and $H \not\subset \operatorname{Ker} \varphi$, then $\operatorname{Res}_H^L \varphi$ has an irreducible component $\theta \neq 1_H$, and so there is a character $\chi \in S$ such that $\varphi \in S(\chi)$. It follows that $\operatorname{Res}_L^G \psi = \beta + \gamma$, where $\beta \in \mathbf{C}[S]$ and the irreducible components of γ have H in their kernels. Since the elements of S vanish on $L - H$, it follows that $\psi(xh) = \gamma(xh) = \gamma(x)$ for $h \in H$. $\square$

(12.5) *Assume Hypothesis* (12.1)*. Let ρ be the mapping defined in Hypothesis* (7.1) *with $A = A(L)$. Let $\psi \in \operatorname{CF}(G)$ be such that ψ is orthogonal to $R(\chi)$ for all $\chi \in S$. Then ψ^{ρ} is constant on $H - H'$.*

Proof. Let $\theta_1, \theta_2 \in \operatorname{Irr} H$ be such that $\theta_i \neq 1_H$ and $\theta_1(1) = \theta_2(1)$. Let $\chi_i = \operatorname{Ind}_H^L \theta_i$. Then $\chi_i \in S$ and, by (12.2) and (5.7), $\{\chi_1, \chi_2, \overline{\chi_1}, \overline{\chi_2}\}$ is coherent. By (5.5), $(\chi_1 - \chi_2)^{\tau} \in \mathbf{Z}[R(\chi_1) \cup R(\chi_2)]$. Thus,

$$(\operatorname{Res}_H^L(\psi^{\rho}), \theta_1 - \theta_2) = (\psi^{\rho}, \chi_1 - \chi_2) = (\psi, (\chi_1 - \chi_2)^{\tau}) = 0.$$

Let $\lambda \in \operatorname{Irr} H'$. If θ_1 and θ_2 are irreducible components of $\operatorname{Ind}_{H'}^H \lambda$, $\theta_1, \theta_2 \neq 1_H$, then $\theta_1(1) = \theta_2(1)$ and $(\operatorname{Ind}_{H'}^H \lambda, \theta_1) = (\operatorname{Ind}_{H'}^H \lambda, \theta_2)$ by (1.7.b). Thus, by the discussion so far, we have $(\operatorname{Res}_H^L(\psi^{\rho}), \theta_1) = (\operatorname{Res}_H^L(\psi^{\rho}), \theta_2)$. It follows that there are numbers $a_{\lambda} \in \mathbf{C}$, for $\lambda \in \operatorname{Irr} H'$, and a number $a \in \mathbf{C}$ such that

$$\operatorname{Res}_H^L(\psi^{\rho}) = \sum_{\lambda \in \operatorname{Irr} H'} a_{\lambda} \operatorname{Ind}_{H'}^H \lambda + a 1_H.$$

For $\lambda \in \operatorname{Irr} H'$, $\operatorname{Ind}_{H'}^H \lambda$ vanishes on $H - H'$; it follows that $\operatorname{Res}_H^L(\psi^{\rho})$ is constant on $H - H'$. $\square$

(12.6) *Assume Hypothesis* (12.1)*. If L is a Frobenius group with kernel H, then $S \subset \operatorname{Irr} L$ and S is coherent.*

Proof. By [Is], Theorem 6.34, $S \subset \operatorname{Irr} L$. If $H^{\#}$ is a TI-subset of G, S is coherent by Theorem (6.8). In case (b) of Definition (8.3) for L, the elements of S are all of the same degree, and so S is coherent by (5.7). Suppose that case (c) of Definition (8.3) holds. By (6.5.b), we may assume that H is a p-group for some prime number p. By (8.2.a), the exponent of L/H is $|L/H|$. Then, by (8.3.c), $|L/H|$ divides $p - 1$, and so S is coherent by (6.5.c). $\square$

(12.7) Theorem. *Every maximal subgroup M of G of Type I is a Frobenius group with kernel M_F.*

Up to the end of the proof of Theorem (12.7), we will assume

(12.8) Hypothesis. *Let π be the set of prime numbers p for which there is a maximal subgroup M of G of Type I such that a Sylow p-subgroup of M/M_F is not cyclic. Suppose that $\pi \neq \emptyset$; let p be the smallest element of π. Let M be a maximal subgroup of G of Type I such that a Sylow p-subgroup of M/M_F is not cyclic. Let $K = M_F$, let $K' = [K, K]$ and let P_0 be a Sylow p-subgroup of M.*

(12.9) *The group P_0 is abelian of rank 2. There is a maximal subgroup L of G such that $P_0 \subset L_s$. There is an element $x \in \Omega_1(P_0)^{\#}$ such that*

$$C_K(x) \not\subset K', \quad N_G(\langle x \rangle) \subset M \quad and \quad C_G(x) \not\subset L.$$

Proof. The first assertion follows from (8.12.a) and Hypothesis (12.8). By (8.17.a), there is a maximal subgroup L of G such that p divides $|L_s|$. By (8.11), L_s contains a Sylow p-subgroup of G. Replacing L by one of its conjugates, we may then assume that $P_0 \subset L_s$. By [BG], Proposition 1.16, there is an element $x \in \Omega_1(P_0)^{\#}$ such that $C_{K/K'}(x) \neq 1$. By [BG], Lemma 1.14, $C_{K/K'}(x) = C_K(x)K'/K'$, and so $C_K(x) \not\subset K'$. By (8.12.b), we then have $N_G(\langle x \rangle) \subset M$ and $C_G(x) \not\subset L$. $\quad\square$

(12.10) *Let $H = L_F$. Then L is a Frobenius group with kernel H.*

Proof. Suppose that L is of type $\mathcal{P}$. If L is of Type II, then, by (8.16), $C_G(y) \subset L$ for all $y \in A(L)$, which contradicts (12.9). By Theorem (10.10) and (11.9.c), L is of Type III and case (b) of (9.7) holds for L. Let U be a complement of H in $[L, L]$. By (11.6), $C_U(H) = 1$, and so U is cyclic by (9.7.b). Since P_0 is not cyclic, $P_0 \subset H$. By (8.6.a), $C_G(y) \subset L$ for all $y \in H^{\#}$, which contradicts (12.9). Thus L is of Type I. By (12.9), $H^{\#}$ is not a TI-subset of G, and so L satisfies condition (b) or condition (c) of Definition (8.3), and $P_0 \subset H$. Let q be a prime divisor of $|L/H|$. In case (8.3.c), q divides $p - 1$. In case (8.3.b), a Sylow p-subgroup P of H is abelian of rank 2. By (8.1.c), there is an element of L of order q which acts fixed-point-freely on $\Omega_1(P)$. Thus q divides $p^2 - 1$, and, consequently, q divides $p - 1$ or $p + 1$. In both cases, we have $q < p$. By the minimality of p, a Sylow q-subgroup of L is cyclic. By (8.2.b), L is then a Frobenius group. $\quad\square$

(12.11) *$M \cap L$ is a complement of K in M and $M \cap L \subset H$.*

Proof. The first assertion follows from (12.9) and (8.13.c1). Let A be a subgroup of $M \cap L$ of order prime to $|H|$. Since H is nilpotent, $P_0 \subset O_p(H)$ and A normalizes $P_0 = O_p(H) \cap M$. By (8.1.c), P_0 does not centralize K, and, by (12.10), if $A \neq 1$, then $P_0 A$ is a Frobenius group with kernel P_0. By (9.1) applied to the action of $P_0 A$ on K, it follows that $C_K(A) \neq 1$. By (12.9), we then obtain $C_K(x) \neq 1$. By (8.1.b), A and x are contained in an abelian subgroup of $M \cap L$, and so A centralizes x. It follows that $A = 1$ and that $M \cap L \subset H$. $\quad\square$

(12.12) *Let E be a complement of H in L and $e = |E|$. Then E is cyclic and e divides $p - 1$ or $p + 1$.*

Proof. Let $P = O_p(H)$ and let x be as in (12.9). Then

$$Z(P) \subset C_P(x) \subset P \cap M = P_0.$$

Since P_0 is of rank 2, $T = \Omega_1 Z(P)$ is elementary abelian of order p or p^2, and E normalizes T. By (12.10), E acts fixed-point-freely on T. Suppose that E normalizes a subgroup of T of order p. Then E can be identified with a subgroup of $\mathrm{Aut}(\mathbf{Z}/p\mathbf{Z})$, whence E is cyclic and e divides $p - 1$. We may then assume that $|T| = p^2$ and that E acts irreducibly on T. By [BG], Theorem 2.6(a), every subgroup of $\mathrm{Aut}((\mathbf{Z}/p\mathbf{Z})^2)$ of order prime to $2p$ is abelian, and so E is abelian. We then see, by applying Schur's Lemma as in the proof of (9.7.b), that $T \rtimes E$ can be identified with $\mathbf{F}_{p^2} \rtimes E_1$, where E_1 is a subgroup of $(\mathbf{F}_{p^2})^*$ acting by multiplication on the additive group $\mathbf{F}_{p^2}$. It follows that E is cyclic and that e divides $p^2 - 1$. Let A be a subgroup of E of order dividing $p - 1$. Then A can be identified with a subgroup of $\mathbf{F}_p^*$, and so A normalizes every subgroup of T of order p. Since $|T| = p^2$, $T = \Omega_1(P_0)$ and so $x \in T$ and A normalizes $\langle x \rangle$. Furthermore, $A \subset M$ by (12.9) and so $A = 1$ by (12.11). Thus e divides $p + 1$. $\qquad\square$

(12.13) Notation. Let $\mathcal{S} = \{\mathrm{Ind}_H^L \theta \mid \theta \in \mathrm{Irr}\, H,\ \theta \neq 1_H\}$ and let τ be the Dade isometry relative to $(A(L), L, G)$. Let τ_1 be an extension of τ to $\mathbf{Z}[\mathcal{S}]$, which exists by (12.6). Let $\chi \in \mathcal{S}$ be such that $\chi(1) = e$ and $\psi = \chi^{\tau_1}$. Let ρ be the mapping defined in Hypothesis (7.1) with $A = A(L)$.

(12.14) *Let x be as in (12.9). If $g \in K$, then $\psi(xg) = \psi^\rho(x) = \chi(x)$.*

Proof. Since p divides $|L_s|$ but not $|M_s|$, L and M are not conjugate. By (5.5), $\psi \in \mathbf{Z}[R(\chi)]$. Thus, by (12.3) and (12.4), ψ is constant on xK. Since $R(x) = C_K(x)$ by Definition (8.14), it follows that $\psi^\rho(x) = \psi(x)$ by definition of ρ. Let $\alpha = \mathrm{Ind}_H^L 1_H - \chi$ and $\mathcal{S} = \{\chi_i \mid 1 \leq i \leq n\}$ with $\chi_1 = \chi$ and $\chi_i(1) = d_i e$. By (12.6), $\mathcal{S} \subset \mathrm{Irr}\, L$. By (7.8.a), there is an integer $a \in \mathbf{Z}$ and a function $\Gamma \in \mathrm{CF}(G)$ orthogonal to $\mathcal{S}^{\tau_1} \cup \{1_G\}$ such that

$$\alpha^\tau = 1_G - \psi + a \sum_{1 \leq i \leq n} d_i \chi_i^{\tau_1} + \Gamma.$$

Then $(a-1)^2 + a^2 \sum_{i>1} d_i^2 = \| -\psi + a \sum d_i \chi_i^{\tau_1} \|^2 \leq \|\alpha^\tau\|^2 - 1 = e$. If $h = |H|$,

$$e^2 \sum_i d_i^2 = \sum_i \chi_i(1)^2 = |L| - |L/H| = e(h - 1),$$

and so $\displaystyle \sum_i d_i^2 = \frac{h - 1}{e}.$ Then

$$a^2 \left(\frac{h - 1}{e} \right) - 2a \leq e - 1.$$

Since $P_0 \subset H$, $p^2 \leq h$, and so $a^2 \left(\dfrac{p^2-1}{e}\right) - 2a \leq e-1$. By (12.12), $e \leq \dfrac{p+1}{2}$,

and so $2(p-1)a^2 - 2a \leq \dfrac{p-1}{2}$. If $f(a) = 2(p-1)a^2 - 2a$, then, because $p \geq 3$,

$f(1) = 2p - 4 > \dfrac{p-1}{2}$. Thus, $f(a) \leq \dfrac{p-1}{2}$ implies that $0 \leq a < 1$. As $a \in \mathbf{Z}$,

$a = 0$. If we set $\chi_0 = \operatorname{Ind}_H^L 1_H$, then

$$((\chi_1 - \chi_0)^\tau, \psi) = (-\alpha^\tau, \psi) = 1$$

and, for $i > 1$,

$$((\chi_i - d_i\chi_0)^\tau, \psi) = ((\chi_i - d_i\chi_1)^\tau - d_i\alpha^\tau, \psi) = 0.$$

By (7.7.a), it follows that $\psi^\rho(x) = \chi_1(x) = \chi(x)$. $\qquad\square$

(12.15) *Let ρ_M be the mapping defined in Hypothesis* (7.1) *with M and $A_1(M)$ in place of L and A. Then $\psi^{\rho_M}(g) = \psi(g)$ for $g \in K^\#$, ψ is constant on $K - K'$ and $\psi(g) \in \mathbf{Z}$ for $g \in K - K'$.*

Proof. Let $g \in K^\#$. If $C_G(g) \subset M$, then $\psi^{\rho_M}(g) = \psi(g)$. Suppose that $C_G(g) \not\subset M$. Let N be a maximal subgroup of G such that $C_G(g) \subset N$. Since M is not a Frobenius group, it follows from (8.13.c4) that N is of Type I. By (8.13.c1), $C_{N_F}(g) \neq 1$, and so N is not a Frobenius group with kernel N_F. Consequently, N is not conjugate to L. By (12.3), (12.4) and (5.5), ψ is constant on gN_F. Then $\psi^{\rho_M}(g) = \psi(g)$ by definition of ρ_M. Since L and M are not conjugate, it then follows from (12.3), (12.5) and (5.5) that ψ is constant on $K - K'$. If $g \in K - K'$, then

$$|K|(\operatorname{Res}_K^G \psi, 1_K) = \sum_{k \in K} \psi(k) = |K'|(\operatorname{Res}_{K'}^G \psi, 1_{K'}) + |K - K'|\psi(g),$$

and so $\psi(g)$ is rational. Since $\psi(g)$ is an algebraic integer, $\psi(g) \in \mathbf{Z}$. $\qquad\square$

(12.16) *Proof of Theorem* (12.7).

If the set π of Hypothesis (12.8) is empty, the theorem holds by (8.2.b). Assume Hypothesis (12.8); we use the notation introduced in items (12.8) to (12.15). By (12.9), there is an element $g \in C_K(x)$ such that $g \notin K'$. Let ε be a primitive p-th root of unity in $\mathbf{C}$. By (1.10.a), $\psi(xg) \equiv \psi(g) \pmod{1 - \varepsilon}$ and $\chi(x) \equiv \chi(1) = e \pmod{1 - \varepsilon}$. Thus, $\psi(g) \equiv e \pmod{1 - \varepsilon}$ by (12.14). By (1.10.b) and (12.15), $\psi(g) \equiv e \pmod{p}$. Also $2e \leq p + 1$ by (12.12), and so $e - p \leq 1 - e$. It follows that $|\psi(g)| \geq e - 1$. By (12.15),

$$\|\psi^{\rho_M}\|^2 \geq \frac{1}{|M|} \sum_{y \in K - K'} |\psi^{\rho_M}(y)|^2 = \frac{|K - K'|}{|M|}|\psi(g)|^2 \geq \frac{|K - K'|}{|M|}(e-1)^2.$$

By (7.8.b), $\|\psi^\rho\|^2 \geq 1 - \dfrac{e}{|H|}$. By Theorem (8.17), $\widetilde{A_1}(M)$ and $\widetilde{A_1}(L)$ are disjoint. By (7.3), we then have

$$1 = \|\psi\|^2 \geq \frac{\psi(1)^2}{|G|} + \frac{1}{|G|} \sum_{g \in \widetilde{A_1}(L) \cup \widetilde{A_1}(M)} |\psi(g)|^2 > \|\psi^{\rho_M}\|^2 + \|\psi^\rho\|^2.$$

Therefore,

$$\frac{|K - K'|}{|M|}(e - 1)^2 + 1 - \frac{e}{|H|} < 1.$$

By (12.11), $|M| = |K||M \cap L| \le |K||H|$. Thus, $\dfrac{|K - K'|}{|K|}(e - 1)^2 < e$. It follows that

$$1 - \frac{|K'|}{|K|} < \frac{e}{(e - 1)^2} = \frac{1}{e}\left(1 + \frac{1}{e - 1}\right)^2 \le \frac{1}{3}\left(1 + \frac{1}{2}\right)^2 = \frac{3}{4},$$

and so $|K/K'| < 4$. This is a contradiction because, by (8.1.c), there is an element of M of order p which acts fixed-point-freely on K/K'. $\qquad\square$

(12.17) *Case* (b) *of Theorem* (8.8) *holds.*

Proof. Suppose that every maximal subgroup of G is of Type I. Let $(L_i)_{1 \le i \le k}$ be a system of conjugacy class representatives of the maximal subgroups of G. Let $H_i = (L_i)_F$. Condition (7.10.a) holds by Theorem (12.7). Since G is simple and L_i is maximal in G, $N_G(H_i) = L_i$. Let $x \in H_i^\#$ be such that $C_G(x) \not\subset L_i$. Let M be a maximal subgroup of G such that $C_G(x) \subset M$. Then, by (8.13.c1), $x \notin M_F$ and $C_{M_F}(x) \ne 1$, and so M is not a Frobenius group. This contradicts Theorem (12.7), and so, by (2.3), $H_i^\#$ is a TI-subset of G and (7.10.b) holds. By (8.17.a), (7.10.c) holds and, by (8.17.c), $G^\# = \bigcup_i (H_i^\#)^G$. Theorem (7.11) shows that this is impossible. $\qquad\square$

13. The Subgroups S and T

Taking (12.17) into account, we will assume until the end

(13.1) Hypothesis. (a) *Let S and T be two maximal subgroups of G which satisfy the conditions of (8.8.b). Let $W = W_1 \times W_2 = S \cap T$. Set $q = |W_1|$ and $p = |W_2|$.*

(b) *Let $P = S_F$ and $Q = T_F$ (the notation M_F is defined before Definition (8.1)). Let U and V be such that $S = (P \rtimes U) \rtimes W_1$ and $T = (Q \rtimes V) \rtimes W_2$. Assume that W_1 normalizes U and that W_2 normalizes V, which is possible by the remark following Definition (8.4). Set $S' = PU$, $T' = QV$, $C = C_U(P)$, $D = C_V(Q)$, $c = |C|$, $d = |D|$, $|U| = uc$ and $|V| = vd$.*

(c) *Let*

$$\mathcal{S} = \{\operatorname{Ind}_{S'}^{S} \theta \mid \theta \in \operatorname{Irr} S',\ P \not\subseteq \operatorname{Ker} \theta\}$$

and

$$\mathcal{T} = \{\operatorname{Ind}_{T'}^{T} \theta \mid \theta \in \operatorname{Irr} T',\ Q \not\subseteq \operatorname{Ker} \theta\}.$$

Denote by τ the Dade isometry relative to $(A_0(S), S, G)$. Also denote by τ the Dade isometry relative to $(A_0(T), T, G)$.

(d) *For $0 \le i < q$, $0 \le j < p$, let ω_{ij} be as in (3.3) and set $\eta_{ij} = \omega_{ij}^{\sigma}$, where σ is as in Theorem (3.2).*

(e) *For $0 \le i < q$, $0 \le j < p$, let μ_{ij} be the characters of S defined in Theorem (4.3) with $L = S$ such that $\operatorname{Ind}_{W}^{S}(\omega_{ij} - \omega_{0j}) = \delta_j(\mu_{ij} - \mu_{0j})$, $\delta_j = \pm 1$. Let ν_{ij} be the characters of T defined in Theorem (4.3) with $L = T$ such that $\operatorname{Ind}_{W}^{T}(\omega_{ij} - \omega_{i0}) = \delta_i'(\nu_{ij} - \nu_{i0})$, $\delta_i' = \pm 1$. Let $\mu_j = \sum_{0 \le i < q} \mu_{ij}$ and $\nu_i = \sum_{0 \le j < p} \nu_{ij}$.*

In this section, S and T play the same role, and results proven for S are valid for T.

(13.2) (a) *S is of Type II or Type III. If $q < p$, then S is of Type II. The group UW_1 is a Frobenius group with abelian kernel U.*

(b) *P is elementary abelian of order p^q.*

(c) *$u \le (p^q - 1)/(p - 1)$.*

(d) *$\mathcal{S}$ is coherent.*

(e) *$A_0(S)$ is a TI-subset of G with normalizer S and the Dade isometry τ relative to $A_0(S)$ coincides with Ind_S^G.*

Proof. By Theorem (10.10) and (11.9.b, c), S is of Type II or III; moreover, if $q < p$, S is of Type II. By the definition of subgroups of Types II and III, U is abelian. By (8.4.d), UW_1 is a Frobenius group. Assertion (b) follows from (10.11) and (11.7). Since

$$(p - 1)^{q-1} = \frac{(p-1)^q}{p - 1} \le \frac{p^q - 1}{p - 1},$$

(9.7) shows that $u \leq (p^q - 1)/(p - 1)$. Assertion (d) follows from (9.11) and from (a) and (b). If $x \in A_0(S)$ and $C_G(x) \not\subset S$, then, by Theorem (8.13), a maximal subgroup of G which contains $C_G(x)$ is a subgroup of Type I which is not a Frobenius group. This is impossible by Theorem (12.7), and so $A_0(S)$ is a TI-subset of G with normalizer S. It follows, therefore, that τ coincides with Ind_S^G. $\qquad\qquad\qquad\qquad\qquad\qquad\qquad\qquad\qquad\qquad\qquad\qquad$ $\square$

In the remainder, we denote by τ_1 an extension of the Dade isometry τ, relative to $A_0(S)$ (relative to $A_0(T)$, respectively), to $\mathbf{Z}[\mathcal{S}]$ (to $\mathbf{Z}[\mathcal{T}]$, respectively).

(13.3) (a) *For $j \geq 1$, μ_j is induced from a linear character of PC and $\mu_j(1) = uq$.*

    (b) *If $\mathcal{S}$ contains no irreducible character of degree uq induced from a linear character of PC, then case (9.7.b) holds for $M = S$, $C = 1$ and $u = (p^q - 1)/(p - 1)$.*

    (c) *The integers δ_j and δ_i' of (13.1.e) are equal to 1. We may assume that either $\mu_j^{\tau_1} = \sum_{0 \leq i < q} \eta_{ij}$ for $j \geq 1$, or $p = 3$ and $\mu_j^{\tau_1} = -\sum_{0 \leq i < q} \eta_{ij'}$ where $\{j, j'\} = \{1, 2\}$.*

Proof. Assertion (a) follows from (9.8.b) and (9.9.a, b) while assertion (b) follows from (9.8.c) and (9.9.a, c). By (a), $\mu_{ij}(1) = u$ for $j \geq 1$. As $(U/C)W_1$ is a Frobenius group, $u \equiv 1 \pmod{q}$. By (4.3.d) and (4.4), it follows that the integers δ_j and δ_i' are equal to 1. By (a), $\mu_j(1)$ does not depend on j for $j \geq 1$. If $\mathcal{S} \cap \mathrm{Irr}\, S = \emptyset$, by Theorem (4.9) we may then set $\mu_j^{\tau_1} = \sum_{0 \leq i < q} \eta_{ij}$ for $j \geq 1$. If $\mathcal{S} \cap \mathrm{Irr}\, S \neq \emptyset$, then, by (5.8), $\mu_j^{\tau_1} = \sum_{0 \leq i < q} \eta_{ij}$, or else $p = 3$ and $\mu_j^{\tau_1} = -\sum_{0 \leq i < q} \eta_{ij'}$, where $j' \neq j$ and $j' \neq 0$. $\qquad\qquad$ $\square$

(13.4) *Suppose that $\mathcal{S}$ contains an irreducible character λ of degree uq induced from a linear character of PC. Then case (9.7.b) holds for $M = T$, $D = 1$ and $v = (q^p - 1)/(q - 1)$.*

Proof. Let $H = PC$ and $K = QD$. If we assume that (13.4) is false, then, by (13.3.b), there is an irreducible character $\theta \in \mathcal{T}$ induced from a linear character of K. Note that $(H^{\#})^G$ and $(K^{\#})^G$ are disjoint. Indeed, let $x \in H^{\#}$ and $g \in G$ be such that $x \in K^g$. Then, by (13.2.e) for T^g, $P \subset C_G(x) \subset T^g$, which is a contradiction. If $\alpha \in \mathbf{Z}[\mathcal{S}, H^{\#}]$ and $\beta \in \mathbf{Z}[\mathcal{T}, K^{\#}]$, then $(\alpha^\tau, \beta^\tau) = 0$ by (13.2.e). Let $\alpha = \lambda - \mu_1$ and $\beta = \theta - \nu_1$. By (13.3.a), $\mathrm{Supp}(\alpha) \subset H^{\#}$ and $\mathrm{Supp}(\beta) \subset K^{\#}$. By (13.3.c), there are indices r and s such that $\mu_1^{\tau_1} = \pm\sum_{0 \leq i < q} \eta_{is}$ and $\nu_1^{\tau_1} = \pm\sum_{0 \leq j < p} \eta_{rj}$. Then

$$(\alpha^\tau, \beta^\tau) = (\lambda^{\tau_1} \pm \sum_i \eta_{is}, \theta^{\tau_1} \pm \sum_j \eta_{rj}) = 0$$

and

$$((\lambda - \bar\lambda)^\tau, (\theta - \bar\theta)^\tau) = (\lambda^{\tau_1} - \bar\lambda^{\tau_1}, \theta^{\tau_1} - \bar\theta^{\tau_1}) = 0.$$

By (4.1) and (5.3.b), the functions η_{ij}, λ^{τ_1} and θ^{τ_1} are pairwise orthogonal. Thus, $0 = (\alpha^\tau, \beta^\tau) = \pm(\eta_{rs}, \eta_{rs})$, which is a contradiction. $\quad\square$

(13.5) *Let $H = PC$ and $\mathcal{S}_1 = \{\operatorname{Ind}_H^S \theta \mid \theta \in \operatorname{Irr} H,\ P \not\subset \operatorname{Ker}\theta\} = \{\zeta_0, \ldots, \zeta_n\}$. Let $\chi \in \mathbf{Z}[\operatorname{Irr} G]$ be such that $((\zeta_i - \zeta_0)^\tau, \chi) = 0$ for $2 \le i \le n$ and let $a = ((\zeta_1 - \zeta_0)^\tau, \chi)$.*

(a) There is an element $\alpha \in \mathbf{Z}[\operatorname{Irr} H]$ such that P is in the kernel of every irreducible component of α and such that, for $x \in H^\#$,

$$\chi(x) = \frac{a}{\|\zeta_1\|^2}\zeta_1(x) + \alpha(x).$$

(b) $\displaystyle \sum_{x \in H^\#} |\chi(x)|^2 = \frac{a^2}{\|\zeta_1\|^2}\left(|S| - \frac{\zeta_1(1)^2}{\|\zeta_1\|^2}\right) - 2a\frac{\zeta_1(1)\alpha(1)}{\|\zeta_1\|^2} + \sum_{x \in H^\#} |\alpha(x)|^2.$

(c) $\displaystyle \sum_{x \in H^\#} |\alpha(x)|^2 \ge (|P| - 1)\alpha(1)^2.$

Proof. (a) By (8.5.a) and (8.6.a), $H^\#$ is a TI-subset of G with normalizer S. Thus Hypothesis (7.1) holds for $L = S$, $A = H^\#$, $\tau = \operatorname{Ind}_S^G$ and $\chi^\rho(x) = \chi(x)$ for $x \in H^\#$. Let $\{\zeta_0, \ldots, \zeta_r\} = \{\operatorname{Ind}_H^S \theta \mid \theta \in \operatorname{Irr} H\}$, where $P \subset \operatorname{Ker}\zeta_i$ for $n < i \le r$. By (13.2.a), H is abelian and so each ζ_i has degree qu. For $i \le n$, $\zeta_i \in \mathbf{Z}[S]$ by (1.5.a), and, by (13.2.e), $(\zeta_i - \zeta_0)^\tau = \operatorname{Ind}_S^G(\zeta_i - \zeta_0)$. By (7.7.a), it follows that, for $x \in H^\#$,

$$\chi(x) = \frac{a}{\|\zeta_1\|^2}\zeta_1(x) + \sum_{n < i \le r} \frac{a_i}{\|\zeta_i\|^2}\zeta_i(x),$$

where $a_i = (\operatorname{Ind}_S^G(\zeta_i - \zeta_0), \chi) \in \mathbf{Z}$. Let $\theta \in \operatorname{Irr} H$ be such that $P \subset \operatorname{Ker}\theta$ and $\zeta = \operatorname{Ind}_H^S \theta$. Then, by (1.5.a, b), $\operatorname{Res}_H^S \zeta = \|\zeta\|^2 \sum_j \theta_j$, where the characters θ_j are the conjugates of θ in S. Thus, for $n < i \le r$, $\dfrac{1}{\|\zeta_i\|^2}\operatorname{Res}_H^S \zeta_i$ is a character of H having P in its kernel.

(b) It is clear that

$$\sum_{x \in H^\#} |\chi(x)|^2$$

$$= \frac{a^2}{\|\zeta_1\|^4}\sum_{x \in H^\#} |\zeta_1(x)|^2 + \frac{a}{\|\zeta_1\|^2}\sum_{x \in H^\#}\left(\zeta_1(x)\overline{\alpha(x)} + \overline{\zeta_1(x)}\alpha(x)\right) + \sum_{x \in H^\#} |\alpha(x)|^2.$$

Since ζ_1 vanishes on $S - H$ and the kernels of irreducible components of ζ_1 do not contain P, $\sum_{x \in H^\#} |\zeta_1(x)|^2 = |S|\,\|\zeta_1\|^2 - \zeta_1(1)^2$ and

$$\sum_{x \in H^\#} \zeta_1(x)\overline{\alpha(x)} = |H|(\operatorname{Res}_H^S \zeta_1, \alpha) - \zeta_1(1)\alpha(1) = -\zeta_1(1)\alpha(1),$$

whence the stated expression follows.

(c) Since the kernel of every irreducible component of α contains P,

$$\sum_{x \in H} |\alpha(x)|^2 = |P| \sum_{x \in C} |\alpha(x)|^2,$$

whence

$$\sum_{x \in H^{\#}} |\alpha(x)|^2 = (|P| - 1)\alpha(1)^2 + |P| \sum_{x \in C^{\#}} |\alpha(x)|^2 \geq (|P| - 1)\alpha(1)^2.$$

$\square$

(13.6) *Suppose that S contains an irreducible character λ of degree uq induced from a linear character of $H = PC$. Then $\sum_{x \in H^{\#}} |\lambda^{\tau_1}(x)|^2 \geq |S| - \lambda(1)^2$.*

Proof. Since S is coherent and $S_1 \subset \mathbf{Z}[S]$, the hypothesis of (13.5) holds with $\zeta_1 = \lambda$, $\chi = \lambda^{\tau_1}$ and $a = 1$. If α is as in (13.5), then

$$\sum_{x \in H^{\#}} |\lambda^{\tau_1}(x)|^2 \geq |S| - \lambda(1)^2 - 2\lambda(1)\alpha(1) + (|P| - 1)\alpha(1)^2.$$

Let $x \in W_2^{\#}$ and $y \in W_1^{\#}$. Then $\lambda(xy) = 0$ and, by (3.2.d), (5.3.b) and (5.5), $\lambda^{\tau_1}(xy) = 0$. Let ε be a primitive qth root of unity in $\mathbf{C}$. Then, by (1.10.a), $\lambda(x) \equiv \lambda^{\tau_1}(x) \equiv 0 \pmod{1 - \varepsilon}$. By (13.5.a), it follows that

$$\alpha(1) = \alpha(x) = \lambda^{\tau_1}(x) - \lambda(x) \equiv 0 \pmod{1 - \varepsilon}.$$

Thus $\alpha(1) \equiv 0 \pmod{q}$ by (1.10.b). If $\alpha(1) = qb$,

$$(|P| - 1)\alpha(1)^2 - 2\lambda(1)\alpha(1) = q^2((|P| - 1)b^2 - 2ub).$$

By (13.2.c), $u \leq (|P| - 1)/2$. Finally, $(|P| - 1)\alpha(1)^2 - 2\lambda(1)\alpha(1) \geq 0$ since b is an integer, and this concludes the proof. $\square$

(13.7) *Let $H = PC$. Then $\sum_{x \in H^{\#}} |\eta_{10}(x)|^2 \geq |H^{\#}|$.*

Proof. By (5.3.b), (5.5) and (13.3.c), η_{10} is orthogonal to S^{τ_1}. As $S_1 \subset \mathbf{Z}[S]$, the hypothesis of (13.5) holds with $\chi = \eta_{10}$ and $a = 0$. Let α be as in (13.5). Let $x \in W_2^{\#}$ and $y \in W_1^{\#}$, and let ε be a primitive qth root of unity in $\mathbf{C}$. By (1.10.a) and (3.2.c), $\eta_{10}(x) \equiv \eta_{10}(xy) = \omega_{10}(y) \equiv 1 \pmod{1 - \varepsilon}$. Therefore, $\alpha(1) = \alpha(x) \equiv 1 \pmod{q}$ by (1.10.b) and (13.5.a), and so $\alpha \neq 0$. Also, $\sum_{x \in H^{\#}} |\alpha(x)|^2 = |H| \, \|\alpha\|^2 - \alpha(1)^2$. By (13.5.a), $\eta_{10}(x) = \alpha(x)$ for $x \in H^{\#}$ and the required inequality then becomes $|H|(\|\alpha\|^2 - 1) \geq \alpha(1)^2 - 1$. By (13.2.a, b), H is abelian. Thus, if $\|\alpha\|^2 = 1$, then $\alpha(1) = \pm 1$ and the required inequality holds. Since $\alpha \neq 0$, we may then assume that $\|\alpha\|^2 \geq 2$. If the required inequality is false, then $\alpha(1)^2 \geq |H|$ and

$$\sum_{x \in H^{\#}} |\eta_{10}(x)|^2 = \sum_{x \in H^{\#}} |\alpha(x)|^2 \geq \sum_{x \in P^{\#}} |\alpha(x)|^2 = (|P| - 1)\alpha(1)^2 > |H|.$$

$\square$

(13.8) *Let $H = PC$. Then $\sum_{x \in H\#} |\eta_{01}(x)|^2 \geq |S'| - u^2$.*

Proof. By (13.3.c), there are integers j and $\delta = \pm 1$ such that $0 < j < p$ and $\mu_j^{\tau_1} = \delta \sum_{0 \leq i < q} \eta_{i1}$. By (13.3.a), μ_j is in the set $\mathcal{S}_1$ of (13.5). The hypothesis of (13.5) has thus been checked with $\zeta_1 = \mu_j$, $\chi = \eta_{01}$ and $a = \delta$. Let α be as in (13.5). By (13.5.b, c),

$$\sum_{x \in H\#} |\eta_{01}(x)|^2 \geq \frac{1}{q}\left(|S| - \frac{(qu)^2}{q}\right) - 2\delta u \alpha(1) + (|P| - 1)\alpha(1)^2.$$

As $u \leq \dfrac{|P| - 1}{2}$ by (13.2.c) and as $\alpha(1) \in \mathbf{Z}$,

$$(|P| - 1)\alpha(1)^2 - 2\delta u \alpha(1) \geq 0,$$

and so $\sum_{x \in H\#} |\eta_{01}(x)|^2 \geq |S'| - u^2$. $\qquad\square$

(13.9) *Let $H = PC$ and $G_0 = G^\# - ((H^\#)^G \cup (Q^\#)^G)$. Let λ be as in (13.6).*

(a) If $x \in G_0$, then $\lambda^{\tau_1}(x) \neq 0$ or $\eta_{10}(x) \neq 0$.

(b) $\sum_{x \in G_0}(|\lambda^{\tau_1}(x)|^2 + |\eta_{10}(x)|^2) \geq |G_0|$.

Proof. (a) By (13.3.c), there are integers $j > 0$ and $\delta = \pm 1$ such that $\mu_j^{\tau_1} = \delta \sum_{0 \leq i < q} \eta_{i1}$. That $(\mu_j - \lambda)^\tau$ vanishes on $G - (H^\#)^G$ follows from (13.2.e) and (13.3.a) . Let $x \in G_0$. Then $\lambda^{\tau_1}(x) = \delta \sum_{0 \leq i < q} \eta_{i1}(x)$. If x is conjugate to an element of $W - (W_1 \cup W_2)$, then $\eta_{10}(x) \neq 0$ by (3.2.c). Suppose that x is not conjugate to an element of $W - (W_1 \cup W_2)$ and that $\lambda^{\tau_1}(x) = \eta_{10}(x) = 0$. By (3.9.b), $\eta_{i0}(x) = 0$ for all $i > 0$. By (3.2.a) and (3.4),

$$\eta_{i1}(x) = \eta_{i0}(x) + \eta_{01}(x) - 1 = \eta_{01}(x) - 1$$

for $i > 0$. Thus $\eta_{i1}(x) = \eta_{11}(x)$ for $i > 0$ and $\eta_{01}(x) = \eta_{11}(x) + 1$. But then $0 = \delta \lambda^{\tau_1}(x) = \sum_{0 \leq i < q} \eta_{i1}(x) = q\eta_{11}(x) + 1$. This is a contradiction because $\eta_{11}(x)$ is an algebraic integer.

(b) Let $A = \{x \in G_0 \mid \lambda^{\tau_1}(x) \neq 0\}$. If $a \in A$ and if $x \in G$ is such that $\langle x \rangle = \langle a \rangle$, then there is an integer k prime to $|\langle a \rangle|$ such that $x = a^k$. Thus, by (1.9.b) applied to the group $\langle a \rangle$, $\lambda^{\tau_1}(x) \neq 0$. Thus A is a union of equivalence classes for the relation $\langle x \rangle = \langle y \rangle$ in G. Now $\sum_{x \in A} |\lambda^{\tau_1}(x)|^2 \geq |A|$ by [Is], Lemma 3.14. Again, if $B = \{x \in G_0 \mid \eta_{10}(x) \neq 0\}$, then $\sum_{x \in B} |\eta_{10}(x)|^2 \geq |B|$. Since $G_0 = A \cup B$, it follows that $\sum_{x \in G_0}(|\lambda^{\tau_1}(x)|^2 + |\eta_{10}(x)|^2) \geq |G_0|$. $\qquad\square$

(13.10) *Suppose that $\mathcal{S}$ contains an irreducible character λ of degree uq induced from a linear character of PC. Let $m = 1 - \dfrac{1}{q-1} - \dfrac{q-1}{q^p} + \dfrac{1}{(q-1)q^p}$.*

Then $\dfrac{u}{c} > \dfrac{mp^{q-1}}{q}$.

Proof. Let H and G_0 be as in (13.9). Since $H^{\#}$ is a TI-subset of G with normalizer S,

$$1 = \frac{1}{|G|} \sum_{x \in G} |\lambda^{\tau_1}(x)|^2 \geq \frac{\lambda^{\tau_1}(1)^2}{|G|} + \frac{1}{|G|} \sum_{x \in G_0} |\lambda^{\tau_1}(x)|^2 + \frac{1}{|S|} \sum_{x \in H^{\#}} |\lambda^{\tau_1}(x)|^2.$$

Thus, by (13.6), we have

$$(13.10.1) \qquad 1 \geq \frac{1}{|G|} + \frac{1}{|G|} \sum_{x \in G_0} |\lambda^{\tau_1}(x)|^2 + 1 - \frac{\lambda(1)^2}{|S|}.$$

Furthermore,

$$1 = \frac{1}{|G|} \sum_{x \in G} |\eta_{10}(x)|^2$$
$$= \frac{\eta_{10}(1)^2}{|G|} + \frac{1}{|G|} \sum_{x \in G_0} |\eta_{10}(x)|^2 + \frac{1}{|S|} \sum_{x \in H^{\#}} |\eta_{10}(x)|^2 + \frac{1}{|T|} \sum_{x \in Q^{\#}} |\eta_{10}(x)|^2.$$

By (13.4), $D = 1$. Then, by (13.7), and (13.8) applied to T, we have

$$(13.10.2) \qquad 1 \geq \frac{1}{|G|} + \frac{1}{|G|} \sum_{x \in G_0} |\eta_{10}(x)|^2 + \frac{|H^{\#}|}{|S|} + \frac{|T'| - v^2}{|T|}.$$

On the other hand, since G is the disjoint union of $\{1\}$, G_0, $(H^{\#})^G$ and $(Q^{\#})^G$, we have

$$(13.10.3) \qquad 1 = \frac{1}{|G|} + \frac{|G_0|}{|G|} + \frac{|H^{\#}|}{|S|} + \frac{|Q^{\#}|}{|T|}.$$

Adding (13.10.1) and (13.10.2), subtracting (13.10.3), and taking (13.9.b) into account, we obtain

$$1 > 1 - \frac{\lambda(1)^2}{|S|} + \frac{|T'| - v^2}{|T|} - \frac{|Q^{\#}|}{|T|},$$

or

$$\frac{\lambda(1)^2}{|S|} > \frac{|T'| - v^2}{|T|} - \frac{|Q^{\#}|}{|T|}.$$

By (13.4), $D = 1$ and $v = (q^p - 1)/(q - 1)$. Then

$$\frac{|T'| - v^2}{|T|} = \frac{1}{p} - \frac{v}{pq^p} = \frac{1}{p} - \frac{1}{p(q-1)} + \frac{1}{p(q-1)q^p}$$

and

$$\frac{|Q^{\#}|}{|T|} = \frac{q^p - 1}{pq^p v} = \frac{q - 1}{pq^p}.$$

It follows that

$$\frac{uq}{cp^q} = \frac{\lambda(1)^2}{|S|} > \frac{1}{p} - \frac{1}{p(q-1)} - \frac{q-1}{pq^p} + \frac{1}{p(q-1)q^p} \,,$$

which gives the inequality of (13.10). □

(13.11) *Under the hypothesis of (13.10), we have:*

(a) *If $q \geq 7$, then $m > \dfrac{8}{10}$.*

(b) *If $q \geq 5$, then $m > \dfrac{7}{10}$.*

(c) *If $q = 3$, then $m > \dfrac{49}{100}$ and $\dfrac{u}{c} > \dfrac{p^2-1}{6}$.*

Proof. It is clear that

$$m > 1 - \frac{1}{q-1} - \frac{q-1}{q^p} > 1 - \frac{1}{q-1} - \frac{1}{q^{p-1}} \geq 1 - \frac{1}{q-1} - \frac{1}{q^2}.$$

If $q \geq 7$, then $m > 1 - \dfrac{1}{6} - \dfrac{1}{49} > \dfrac{8}{10}$. If $q \geq 5$, then $m > 1 - \dfrac{1}{4} - \dfrac{1}{25} > \dfrac{7}{10}$.
Suppose that $q = 3$. Then

$$m = 1 - \frac{1}{2} - \frac{2}{3^p} + \frac{1}{2 \cdot 3^p} = \frac{1}{2} - \frac{3}{2 \cdot 3^p} \geq \frac{1}{2} - \frac{1}{2 \cdot 3^4} > \frac{49}{100}.$$

By (13.10), $\dfrac{u}{c} > \dfrac{mp^2}{3} = \dfrac{p^2 - f(p)}{6}$, where $f(x) = \dfrac{x^2}{3^{x-1}}$. But

$$\frac{f(x+1)}{f(x)} = \frac{1}{3}(1 + \frac{1}{x})^2 \leq \frac{16}{27} < 1$$

for $x \geq 3$, and so $f(p) \leq f(3) = 1$. □

(13.12) $c = 1$.

Proof. By (13.3.b), we may assume that the hypothesis of (13.10) holds. By (13.2.c) and (13.10),

$$m < \frac{uq}{cp^{q-1}} \leq \frac{q}{cp^{q-1}} \frac{(p^q - 1)}{(p-1)}.$$

Suppose that $c \neq 1$. Since W_1 acts fixed-point-freely on C and c is odd, $c \geq 2q + 1$. Thus

$$m < \frac{q(p^q - 1)}{(2q+1)p^{q-1}(p-1)} < \frac{qp}{(2q+1)(p-1)}.$$

Assume first that $p = 3$. Then $m < \dfrac{3q}{4q+2} < \dfrac{3}{4} < \dfrac{8}{10}$. By (13.11.a), $q = 5$.
We then obtain $m < \dfrac{15}{22} < \dfrac{7}{10}$, which contradicts (13.11.b).

Suppose that $p \geq 5$. Then

$$m < \frac{p}{2(p-1)} = \frac{1}{2}\Big(1 + \frac{1}{p-1}\Big) \leq \frac{1}{2}\Big(1 + \frac{1}{4}\Big) < \frac{7}{10}.$$

By (13.11.b), $q = 3$. Since W_1 acts fixed-point-freely on C, $c = 7$ or $c \geq 13$. If $c \geq 13$, then

$$m < \frac{3(p^3 - 1)}{13p^2(p-1)} = \frac{3}{13}\Big(1 + \frac{1}{p} + \frac{1}{p^2}\Big) \leq \frac{3 \cdot 31}{13 \cdot 25} < \frac{49}{100}.$$

Thus, by (13.11.c), we have $c = 7$. If $p \geq 11$, then

$$m < \frac{3(p^3 - 1)}{7p^2(p-1)} \leq \frac{3 \cdot (11^2 + 11 + 1)}{7 \cdot 11^2} = \frac{399}{847} < \frac{49}{100}.$$

Thus, $p < 11$ by (13.11.c). As p is prime to c, $p = 5$ and $c = 7$. As $p-1$ has no odd divisor $\neq 1$, case (9.7.b) holds for $M = S$. Thus u divides $(p^q - 1)/(p-1) = 31$. Then c is prime to u, and so PC is a normal nilpotent Hall subgroup of S. This is a contradiction because $P = S_F$. $\qquad\square$

(13.13) *Suppose that case (9.7.a) holds for $M = S$. Then $q = 3$ and $u = (p-1)^2/4$.*

Proof. By (13.3.b), the hypothesis of (13.10) holds. Since u is odd, (9.7.a) implies that

$$u \leq \Big(\frac{p-1}{2}\Big)^{q-1} < \frac{p^{q-1}}{2^{q-1}}.$$

Then, by (13.10) and (13.12),

$$m < \frac{qu}{p^{q-1}} < \frac{q}{2^{q-1}}.$$

With $f(x) = \dfrac{x}{2^{x-1}}$, $\dfrac{f(x+1)}{f(x)} = \dfrac{x+1}{2x} \leq 1$ for $x \geq 1$. Thus, if $q \geq 5$, then $m < f(5) = 5/16 < 7/10$. By (13.11.b), $q = 3$; also, u divides $(p-1)^2/4$. If $u \neq (p-1)^2/4$, then $u \leq (p-1)^2/8$. This implies that $(p^2-1)/6 < (p-1)^2/8$ by (13.11.c), and so $4(p+1) < 3(p-1)$, which is a contradiction. $\qquad\square$

(13.14) *The number $(p^q - 1)/(p-1)$ is odd. If $p \equiv 1 \pmod{q}$, then q divides $(p^q - 1)/(p-1)$. If $p \not\equiv 1 \pmod{q}$, then $(p^q - 1)/(p-1)$ is prime to $(p-1)$ and, if $x \in \mathbf{N}$ is a divisor of $(p^q - 1)/(p-1)$, then $x \equiv 1 \pmod{q}$.*

Proof. Let r be a prime divisor of $p - 1$. Then

$$\frac{p^q - 1}{p - 1} = 1 + p + \cdots + p^{q-1} \equiv 1 + 1 + \cdots + 1 = q \pmod{r}.$$

Taking $r = 2$, we see that $(p^q - 1)/(p-1)$ is odd. Taking $r = q$, we see that, if $p \equiv 1 \pmod{q}$, then q divides $(p^q - 1)/(p-1)$.

Suppose that $p \not\equiv 1 \pmod{q}$. For every prime divisor r of $p-1$,

$$\frac{p^q - 1}{p - 1} \equiv q \not\equiv 0 \pmod{r},$$

and so $(p^q-1)/(p-1)$ is prime to $p-1$. Let $x \in \mathbf{N}$ be a divisor of $(p^q-1)/(p-1)$. To show that $x \equiv 1 \pmod{q}$, we may assume that x is a prime number. As $(p^q - 1)/(p - 1)$ is prime to $p - 1$, x does not divide $p - 1$. It follows that the image of p in $(\mathbf{Z}/x\mathbf{Z})^*$ has order q, and so q divides $|(\mathbf{Z}/x\mathbf{Z})^*| = x - 1$. $\quad\square$

(13.15) *Suppose that case* (9.7.b) *holds with* $M = S$. *Then*

$$u = \begin{cases} \dfrac{p^q - 1}{p - 1} & \text{if } p \not\equiv 1 \pmod{q}, \\[2mm] \dfrac{p^q - 1}{q(p - 1)} & \text{if } p \equiv 1 \pmod{q}. \end{cases}$$

Proof. Let x be the integer for which $ux = (p^q-1)/(p-1)$. If $p \equiv 1 \pmod{q}$, then (13.14) implies that q divides x since u is prime to q. If $x \neq q$, then $x \geq 3q$. If $p \not\equiv 1 \pmod{q}$, then $x \equiv 1 \pmod{q}$ by (13.14), and so $x \geq 2q+1$ if $x \neq 1$. We may then assume that $x \geq 2q+1$. By (13.3.b), there is a character λ satisfying the hypotheses of (13.10). Then, by (13.10) and (13.12),

$$m < \frac{q(p^q - 1)}{p^{q-1}x(p - 1)} \leq \frac{q(p^q - 1)}{(2q + 1)p^{q-1}(p - 1)} < \frac{qp}{(2q + 1)(p - 1)}.$$

Assume first that $p = 3$. Then $m < \dfrac{3}{4} < \dfrac{8}{10}$. By (13.11.a), $q = 5$. We then obtain $m < \dfrac{15}{22} < \dfrac{7}{10}$, which contradicts (13.11.b).

Suppose that $p \geq 5$. Then $m < \dfrac{p}{2(p - 1)} \leq \dfrac{5}{8} < \dfrac{7}{10}$. By (13.11.b), it follows that $q = 3$. By (13.11.c),

$$\frac{p^2 - 1}{6} < u = \frac{p^2 + p + 1}{x} \leq \frac{p^2 + p + 1}{7}.$$

Thus $p^2 - 6p - 13 < 0$. Consequently, $p = 5$ or $p = 7$. But x divides $p^2 + p + 1$ which is 31 or $57 = 3 \cdot 19$. As $u = (p^2 + p + 1)/x \neq 1$ and u is prime to 3, we see that this is impossible if $x \geq 2q + 1 = 7$. $\quad\square$

(13.16) $N_G(W_1) = C_G(W_1) = QW_2$.

Proof. Since $Q^{\#}$ is a TI-subset of G with normalizer T, $N_G(W_1) = N_T(W_1)$. As $QW_2 \subset C_G(W_1)$, it follows that $N_G(W_1) = QN_V(W_1)W_2$. Let $K = N_V(W_1)$. By Maschke's Theorem, there is a subgroup Q_1 of Q such that $Q = W_1 \times Q_1$ and such that KW_2 normalizes Q_1. If $K \neq 1$, then KW_2 is a Frobenius group with kernel K and $C_{Q_1}(W_2) = W_1 \cap W_2 = 1$. By (9.1), it follows that K centralizes Q_1. The group $C_{KW_2}(W_1)$ is a normal subgroup of KW_2 which contains W_2. Then, by [BG], Lemma 3.2, $K \subset C_{KW_2}(W_1)$. Thus K centralizes Q and, by (13.12), $K = 1$.   □

(13.17) *Suppose that S is of Type II. Let L be a maximal subgroup of G such that $N_G(U) \subset L$ and $H = L_F$.*

   (a) *L is a Frobenius group with kernel H.*

   (b) *$U \subset H$.*

   (c) *Either $L = H \rtimes W_1$, or else there is an element $y \in Q$ such that $L = H \rtimes (W_1 W_2^y)$.*

Proof. (a) Suppose that there is an element $g \in G$ such that $L^g = S$. Since S is solvable and U is a Hall subgroup of S, there is an element $x \in S$ such that $U^{gx} = U$. Then $N_G(U) = N_G(U)^{gx} \subset L^{gx} = S$, which contradicts the definition of subgroup of Type II. Suppose that L is conjugate to T. Then $|H| = q^p$. Thus, as $W_1 \subset N_G(U) \subset L$, $W_1 \subset H$. Since $U \subset L$ normalizes H and u is prime to q, $[U, W_1] \subset H \cap U = 1$. This contradicts (13.2.a). By (8.8.b4), L is thus of Type I and, by Theorem (12.7), L is a Frobenius group.

(b) Since q divides $|T_s|$, (a) and (8.17.a) show that $|H|$ is prime to q. Thus $W_1 \cap H = 1$. Suppose that $U \cap H = 1$. Then UW_1 is a Frobenius group contained in L and UW_1 acts fixed-point-freely on H. By (9.1), $|H| = 1$, which is a contradiction. Thus $U \cap H \neq 1$ and $U \subset C_L(U \cap H) \subset H$.

(c) We have seen that $W_1 \cap H = 1$. Let E be a complement to H in L such that $W_1 \subset E$. By [H], Kapitel V, Satz 8.18 b), we know that, in a Frobenius complement of odd order, every subgroup of prime order is normal. Thus $E \subset N_G(W_1)$. By (13.16), $W_1 \subset E \subset QW_2$. By [BG], Proposition 3.9, the Sylow subgroups of E are cyclic. Thus $E = W_1$ or $|E| = pq$. In the second case, it follows from Sylow's Theorem that there is an element $y \in Q$ such that $E = W_1 W_2^y$.   □

(13.18) *Let j be such that $0 < j < p$ and let $\beta_j = \operatorname{Ind}_{PW_1}^S 1_{PW_1} - \mu_{0j}$.*

   (a) $\operatorname{Supp}(\beta_j) \subset P^\# \cup (W - (W_1 \cup W_2))^S \subset A_0(S)$.

   (b) $\|\beta_j\|^2 = \dfrac{u-1}{q} + 2.$

   (c) *$\Gamma = \beta_j^\tau - 1_G + \eta_{0j}$ is independent of j, orthogonal to 1_G and real.*

   (d) *Set $\Gamma = X + Y$ where X is a linear combination of the functions η_{ik} and Y is orthogonal to the functions η_{ik} $(0 \le i < q,\ 0 \le k < p)$. Then*
$$\|Y\|^2 \le \frac{u-1}{q}.$$

Proof. (a) By (4.5.a), $\operatorname{Res}_{S'}^S \mu_{0j} = (1/q)\operatorname{Res}_{S'}^S \mu_j$, and so μ_{0j} vanishes on $S' - P$ by (13.3.a) and (13.12). Also, $\mu_{0j}(1) = (1/q)\mu_j(1) = u$, and so $\beta_j(1) = 0$. As $S' \cap (PW_1)^S = P$, $\operatorname{Supp}(\beta_j) \cap S' \subset P^\#$. Let $x \in W_1^\#$. Then, by (4.3.c) and (13.3.c), $\mu_{0j}(x) = \omega_{0j}(x) = 1$. Let $\overline{S} = S/P$, $\overline{W_1} = W_1 P/P$ and $\overline{U} = UP/P$.

By (1.6.b), $\mathrm{Ind}_{PW_1}^{S} 1_{PW_1}$ can be identified with $\gamma = \mathrm{Ind}_{\overline{W_1}}^{\overline{S}} 1_{\overline{W_1}}$. Since $\overline{S} = \overline{U W_1}$ is a Frobenius group, if $x \in \overline{W_1}^{\#}$ and $y \in \overline{U}$, $x^y \in \overline{W_1}$ implies that $y = 1$. Then, by definition of induction, $\gamma(x) = 1$. Thus, if $x \in W_1^{\#}$, $\beta_j(x) = 0$. If $a \in S - S'$, then, by (2.1), there is an element $x \in W_1^{\#}$ such that a is conjugate in S to an element of xW_2. Thus, if $a \in \mathrm{Supp}(\beta_j)$, a is conjugate to an element of $W - (W_1 \cup W_2)$.

(b) Since $\mu_j \in \mathcal{S}$ and $\mathrm{Res}_{S'}^{S} \mu_{0j} = (1/q)\mathrm{Res}_{S'}^{S} \mu_j$, we have $P \not\subset \mathrm{Ker}\,\mu_{0j}$, and so $\mathrm{Ind}_{PW_1}^{S} 1_{PW_1}$ is orthogonal to μ_{0j}. Thus $\|\beta_j\|^2 = \|\mathrm{Ind}_{PW_1}^{S} 1_{PW_1}\|^2 + 1$. Let γ, $\overline{S}$, $\overline{W_1}$ and $\overline{U}$ be as in the proof of (a). Then $\gamma(1) = u$ and $\gamma(x) = 0$ for $x \in \overline{U}^{\#}$. Since $\overline{S}$ is a Frobenius group, $\overline{S} - \overline{U}$ is the union of the conjugates of $\overline{W_1}^{\#}$, and so, for $x \in \overline{S} - \overline{U}$, $\gamma(x) = 1$. Thus

$$\|\gamma\|^2 = \frac{1}{uq}(u^2 + (q-1)u) = \frac{u-1}{q} + 1.$$

(c) Let $\Gamma_j = \beta_j^{\tau} - 1_G + \eta_{0j}$. By Frobenius reciprocity,

$$(\beta_j^{\tau}, 1_G) = (\beta_j, 1_S) = (\mathrm{Ind}_{PW_1}^{S} 1_{PW_1}, 1_S) = (1_{PW_1}, 1_{PW_1}) = 1.$$

Thus $(\Gamma_j, 1_G) = 0$. Let k be such that $0 < k < p$. By (4.8) and (13.3.c),

$$\Gamma_j - \Gamma_k = (\beta_j - \beta_k)^{\tau} - (\eta_{0k} - \eta_{0j}) = (\mu_{0k} - \mu_{0j})^{\tau} - (\eta_{0k} - \eta_{0j}) = 0.$$

Let k be such that $\overline{\omega_{0j}} = \omega_{0k}$. By (3.9.a) and (4.3.b), $\overline{\mu_{0j}} = \mu_{0k}$ and $\overline{\eta_{0j}} = \eta_{0k}$. Thus, $\overline{\Gamma_j} = \overline{\beta_j}^{\tau} - 1_G + \overline{\eta_{0j}} = \beta_k^{\tau} - 1_G + \eta_{0k} = \Gamma_k = \Gamma_j$.

(d) By (b),

$$\frac{u-1}{q} + 2 = \|\beta_j^{\tau}\|^2 = \|1_G - \eta_{0j} + X\|^2 + \|Y\|^2.$$

Let $a = (X, \eta_{0j})$. By (c), $(X, \overline{\eta_{0j}}) = a$, and so

$$\|1_G - \eta_{0j} + X\|^2 \geq 1 + (a-1)^2 + a^2 \geq 2,$$

whence $\|Y\|^2 \leq \dfrac{u-1}{q}$. $\qquad\qquad\square$

(13.19) *Let L be a maximal subgroup of G of Type I, $H = L_F$ and $e = |L : H|$. Let $\mathcal{L} = \{\mathrm{Ind}_H^{L} \theta \mid \theta \in \mathrm{Irr}\, H,\ \theta \neq 1_H\}$ and let τ be the Dade isometry relative to $(A(L), L, G)$. There is an extension τ_1 of τ to $\mathbf{Z}[\mathcal{L}]$ and there is a character $\varphi \in \mathcal{L}$ such that $\varphi(1) = e$. Let $\beta_L = \mathrm{Ind}_H^{L} 1_H - \varphi$ and $\beta_S = \mathrm{Ind}_{PW_1}^{S} 1_{PW_1} - \mu_{01}$.*

(a) $\tilde{A}(L) \cap (P^G \cup W^G) = \emptyset$.

(b) $\mathcal{L}^{\tau_1}$ *is orthogonal to* η_{ij} *for* $0 \leq i < q$, $0 \leq j < p$.

(c) $(\beta_L^\tau, \eta_{0j})$ *is independent of j for $1 \leq j < p$ and one of the following two cases holds.*

(c1) $(\beta_S^\tau, \varphi^{\tau_1}) \equiv 1 \pmod 2$ *and* $\dfrac{|H| - 1}{e} \leq \dfrac{u - 1}{q}$.

(c2) $(\beta_L^\tau, \eta_{0j}) \equiv 1 \pmod 2$ *for $j \geq 1$, and $p \leq e$.*

Proof. The existence of τ_1 follows from (12.6) and Theorem (12.7); also, $\mathcal{L} \subset \mathrm{Irr}\, L$. The existence of φ is clear. Since L is conjugate neither to S nor to T, (8.17.a) shows that $|H|$ is prime to pq. As the order of an element of $\tilde{A}(L)$ is divisible by a prime divisor of $|H|$, assertion (a) follows. If $\psi \in \mathcal{L}$, $(\psi - \overline{\psi})^\tau$ vanishes on $W - (W_1 \cup W_2)$ by (a) and the definition of τ. In the notation of Hypothesis (3.6), $NC((\psi - \overline{\psi})^\tau) \leq \|\psi - \overline{\psi}\|^2 = 2$, and so, by (3.8), ψ^{τ_1} is orthogonal to all the η_{ij}. By (13.18.a),

$$\mathrm{Supp}(\mu_{0j} - \mu_{01}) \subset P^{\#} \cup (W - (W_1 \cup W_2))^S$$

for $j \geq 1$. Then, by (4.8) and (a), $(\beta_L^\tau, \eta_{0j} - \eta_{01}) = (\beta_L^\tau, (\mu_{0j} - \mu_{01})^\tau) = 0$ for $j \geq 1$. Set $\beta_L^\tau = 1_G - \varphi^{\tau_1} + \Gamma_L$ and $\beta_S^\tau = 1_G - \eta_{01} + \Gamma_S$. By (5.9), $\beta_L^\tau - \overline{\beta_L^\tau} = (\overline{\varphi} - \varphi)^\tau = \overline{\varphi^{\tau_1}} - \varphi^{\tau_1}$, and so Γ_L is real. By (7.8.a), $(\Gamma_L, 1_G) = 0$. By (13.18.c) and (1.1), it follows that (Γ_L, Γ_S) is even. By (a) and (13.18.a),

$$0 = (\beta_L^\tau, \beta_S^\tau) = 1 - (\varphi^{\tau_1}, \Gamma_S) - (\Gamma_L, \eta_{01}) + (\Gamma_L, \Gamma_S).$$

Thus $(\Gamma_S, \varphi^{\tau_1}) + (\Gamma_L, \eta_{01}) \equiv 1 \pmod 2$. It follows that

$$(\beta_S^\tau, \varphi^{\tau_1}) = (\Gamma_S, \varphi^{\tau_1}) \equiv 1 \pmod 2$$

or

$$(\beta_L^\tau, \eta_{01}) = (\Gamma_L, \eta_{01}) \equiv 1 \pmod 2.$$

Suppose that $a = (\beta_S^\tau, \varphi^{\tau_1}) \equiv 1 \pmod 2$. Let $\mathcal{L} = \{\varphi_i \mid 1 \leq i \leq n\}$ where $\varphi_1 = \varphi$ and $\varphi_i(1) = a_i e$. By (a) and (13.18.a), $((\varphi_i - a_i\varphi)^\tau, \beta_S^\tau) = 0$ for $1 \leq i \leq n$. Thus $\Gamma_S = a \sum_{i=1}^{n} a_i \varphi_i^{\tau_1} + \Delta$, where Δ is orthogonal to $\mathcal{L}^{\tau_1}$. Since $a \neq 0$, it follows from (13.18.d) that $\sum_{i=1}^{n} a_i^2 \leq (u - 1)/q$. As

$$\sum a_i^2 = \frac{|L| - |L : H|}{e^2} = \frac{|H| - 1}{e},$$

it follows that

$$\frac{|H| - 1}{e} \leq \frac{u - 1}{q}.$$

Suppose that $(\beta_L^\tau, \eta_{01}) \equiv 1 \pmod 2$. Then $(\beta_L^\tau, \eta_{0j}) \neq 0$ for $1 \leq j \leq p$, and so, by (7.8.b), $p - 1 \leq e - 1$. $\qquad \square$

14. Non-existence of G

In this section, we continue to assume Hypothesis (13.1) and we also assume

(14.1) Hypothesis. $q < p$.

We will prove the following theorem.

(14.2) Theorem. (a) *Let $F = \mathbf{F}_{p^q}$ and let U^* be the subgroup of the multiplicative group F^* which has order $(p^q - 1)/(p - 1)$. There is an isomorphism from PU to $F \rtimes U^*$, where U^* acts on the additive group F by multiplication. This isomorphism identifies P with the additive group F, U with U^* and W_2 with the additive subgroup $\mathbf{F}_p$ of F. Furthermore, $(p^q - 1)/(p - 1)$ is prime to $p - 1$.*

(b) *Q is an elementary abelian q-group, W_2 normalizes Q and there is an element $y \in Q$ such that W_2^y normalizes U.*

Theorem C of [BG], Appendix C, shows that statements (a) and (b) of Theorem (14.2) imply that $p \leq q$, which contradicts Hypothesis (14.1). Consequently, the proof of Theorem (14.2) will show that G does not exist and will complete the proof of the Feit-Thompson Theorem.

By (13.2.a), S is of Type II. By (13.17) and (12.6), we may then assume

(14.3) Hypothesis. (a) *L is a maximal subgroup of G which contains $N_G(U)$, and $H = L_F$. Set $\mathcal{L} = \{\mathrm{Ind}_H^L \theta \mid \theta \in \mathrm{Irr}\, H,\ \theta \neq 1_H\}$. Denote by τ the Dade isometry relative to $(A(L), L, G)$, by τ_1 an extension of τ to $\mathbf{Z}[\mathcal{L}]$ and by φ a character of $\mathcal{L}$ of degree $|L : H|$.*

(b) *Set*

$$\begin{aligned}
\beta_S &= \mathrm{Ind}_{PW_1}^S 1_{PW_1} - \mu_{01}, \\
\beta_T &= \mathrm{Ind}_{QW_2}^T 1_{QW_2} - \nu_{10}
\end{aligned}$$

and

$$\beta_L = \mathrm{Ind}_H^L 1_H - \varphi.$$

(14.4) *Case (9.7.b) holds for $M = T$, and $v = (q^p - 1)/(q - 1)$.*

Proof. Since $q < p$, $p \neq 3$, and so, by (13.13), case (9.7.b) holds for $M = T$. Moreover, $q \not\equiv 1 \pmod{p}$, and so $v = (q^p - 1)/(q - 1)$ by (13.15). $\qquad\square$

(14.5) *There is an element $y \in Q$ such that $L = H \rtimes (W_1 W_2^y)$.*

Proof. If (14.5) is false, then, by (13.17.c), $|L : H| = q < p$. Thus case (c1) of (13.19) holds. As $U \subset H$ by (13.17.b), it follows that

$$\frac{u - 1}{q} \leq \frac{|H| - 1}{q} \leq \frac{u - 1}{q}.$$

Thus $H = U$. But then $N_G(U) \subset L = UW_1 \subset S$, which contradicts the fact that S is of Type II. $\qquad\square$

(14.6) *Case (9.7.b) holds for $M = S$.*

Proof. Suppose that case (9.7.a) holds for $M = S$. By (13.13), $q = 3$ and $u = (p-1)^2/4$. Let r be a prime divisor of u, R_0 a Sylow r-subgroup of U and R a Sylow r-subgroup of H such that $R_0 \subset R$. By (9.7.a), U is isomorphic to a subgroup of the product of two cyclic groups of order dividing $(p-1)/2$. As $u = (p-1)^2/4$, U is then the product of two cyclic groups of order $(p-1)/2$. Thus R_0 is of rank 2 and, by [BG], Proposition 1.16, there is an element $x \in R_0^\#$ such that $C_P(x) \neq 1$. Then, by (13.2.e), $Z(R) \subset C_G(x) \subset S$, and so $Z(R) \subset R_0$. Consequently, $\Omega_1 Z(R)$ is elementary abelian of order r or r^2. If y is as in (14.5), then W_2^y acts fixed-point-freely on $\Omega_1 Z(R)$ by (13.17). Thus p divides $r^2 - 1$ and so p divides $r - 1$ or $r + 1$. But then $p < r$, which is a contradiction since r divides $(p-1)/2$. $\qquad\square$

(14.7) *If U is characteristic in H, then Theorem (14.2) holds.*

Proof. By (13.2.b) and (14.5), assertion (b) of Theorem (14.2) holds if U is characteristic in H. By (14.6), (9.7) and (13.12), it then suffices to show that $u = (p^q - 1)/(p - 1)$. If this is not the case, then $qu = (p^q - 1)/(p - 1)$ by (13.15). As W_2^y acts fixed-point-freely on U,
$$q \equiv qu = 1 + p + \cdots + p^{q-1} \equiv 1 \pmod{p}.$$
This is a contradiction because $q < p$. $\qquad\square$

(14.8) (a) $q^{p+1} > p^{q+1}$.

  (b) $\dfrac{v-1}{p} > \dfrac{u-1}{q}$.

Proof. (a) Let $f(x) = (\log x)/(x + 1)$. Inequality (a) is equivalent to the inequality $f(p) < f(q)$. But $(x + 1)^2 f'(x) = 1 + 1/x - \log x$. For $x \geq 5$, $(x+1)^2 f'(x) \leq 1 + 1/5 - \log 5 < 0$, and so f is strictly decreasing for $x \geq 5$. If $q \geq 5$, then $f(p) < f(q)$. If $q = 3$, then $p \geq 5$ and $f(p) \leq f(5)$. But $5^2 < 3^3$ and so $5^4 < 3^6$ and $f(5) < f(3)$.

  (b) By (13.2.c) and (14.4), $u \leq (p^q - 1)/(p - 1)$ and $v = (q^p - 1)/(q - 1)$. It then suffices to show that
$$\frac{p^q - p}{q(p-1)} < \frac{q^p - q}{p(q-1)}.$$

By (a) and Hypothesis (14.1), $0 < p^{q+1} - p^2 < q^{p+1} - q^2$, whence
$$(q-1)(p^{q+1} - p^2) < (p-1)(q^{p+1} - q^2)$$
and
$$\frac{p^q - p}{q(p-1)} < \frac{q^p - q}{p(q-1)}.$$

$\qquad\square$

(14.9) *T is of Type II.*

Proof. Suppose that T is of Type III. Let

$$\mathcal{T}_1 = \{\operatorname{Ind}_{T'}^{T} \theta \mid \theta \in \operatorname{Irr} T', \ \theta \neq 1_{T'}, \ Q \subset \operatorname{Ker}\theta\}$$

and let $\zeta \in \mathcal{T}_1$. Let τ_1 be an extension to $\mathbf{Z}[\mathcal{T}_1]$ of the Dade isometry τ for T, which exists by (5.7). By (5.3.b) and (5.5), $\mathcal{T}_1^{\tau_1}$ is orthogonal to all the η_{ij}. Set $(\nu_0 - \zeta)^{\tau} = 1_G - \zeta^{\tau_1} + \Delta$ and $\beta_S^{\tau} = 1_G - \eta_{01} + \Gamma$. By (5.9), $\overline{(\nu_0 - \zeta)^{\tau}} - (\nu_0 - \zeta)^{\tau} = (\zeta - \overline{\zeta})^{\tau} = \zeta^{\tau_1} - \overline{\zeta^{\tau_1}}$, and so Δ is real. By (11.9.a), Δ is orthogonal to 1_G and to η_{01}, and, by (13.18.c), Γ is real and orthogonal to 1_G. By (13.18.a) and (13.2.e), $\operatorname{Supp}(\beta_S^{\tau}) \cap \tilde{A}(T) = \emptyset$, and so

$$0 = ((\nu_0 - \zeta)^{\tau}, \beta_S^{\tau}) = 1 - (\zeta^{\tau_1}, \Gamma) + (\Delta, \Gamma).$$

As Δ and Γ are real and orthogonal to 1_G, $(\zeta^{\tau_1}, \Gamma) \equiv 1 \pmod 2$. Since $(QV/Q)W_2$ is a Frobenius group with abelian kernel $\cong V$, $|\mathcal{T}_1| = (v-1)/p$. By (13.18.d), it follows that $\dfrac{v-1}{p} \leq \dfrac{u-1}{q}$. This contradicts (14.8). $\qquad\square$

Taking (14.9), (13.17) and (12.6) into account, we will assume

(14.10) Hypothesis. *Let M be a maximal subgroup of G which contains $N_G(V)$ and let $K = M_F$. Set $\mathcal{M} = \{\operatorname{Ind}_K^M \theta \mid \theta \in \operatorname{Irr} K, \ \theta \neq 1_K\}$. Denote by τ the Dade isometry relative to $(A(M), M, G)$, by τ_1 an extension of τ to $\mathbf{Z}[\mathcal{M}]$ and by ψ a character in $\mathcal{M}$ of degree $|M : K|$. Set $\beta_M = \operatorname{Ind}_K^M 1_K - \psi$.*

(14.11) $K = V$ *and* $|M : K| = pq$.

Proof. Since, by (14.9), $N_G(V) \not\subset T$, by (13.17.c) it suffices to show that $K = V$. Let $k = |K|$ and $e = |M : K|$, and suppose that $V \neq K$.

(14.11.1) *We have $k > 2pv$ and $\dfrac{k-1}{e} \geq \dfrac{v-1}{p} > \dfrac{u-1}{q}$.*

Proof. Set $k = vx$. By (13.17), x is an integer and $e \leq pq$. Since W_2 acts fixed-point-freely on K and on V, $vx \equiv x \equiv 1 \pmod p$. Then, since $x \neq 1$, $x \geq 2p + 1$. Thus $k > 2pv > qv$ and $\dfrac{k-1}{e} > \dfrac{k-1}{pq} > \dfrac{qv-1}{pq} > \dfrac{v-1}{p}$. The conclusion then follows from (14.8). $\qquad\square$

(14.11.2) *We have $e = pq$ and $\beta_M^{\tau} = \sum_{0 \leq i < q, 0 \leq j < p}(\pm\eta_{ij}) - \chi$, where $\chi = \psi^{\tau_1}$ or $-\overline{\psi}^{\tau_1}$.*

Proof. Let $a_{ij} = (\beta_M^{\tau}, \eta_{ij})$. By (14.11.1) and (13.19.c), $a_{0j} \equiv a_{i0} \equiv 1 \pmod 2$ for $i \geq 1$, $j \geq 1$. By (7.8.a), $a_{00} = 1$ and, by (13.19.a), β_M^{τ} vanishes on $W - (W_1 \cup W_2)$. By (3.7), $a_{ij} = a_{i0} + a_{0j} - a_{00} \equiv 1 \pmod 2$ for all i, j. By (7.8.b) and (13.19.b),

$$pq - 1 \leq \sum_{(i,j)\neq(0,0)} a_{ij}^2 \leq e - 1.$$

As $e \le pq$, we see that $e = pq$. Moreover, $a_{ij}^2 = 1$ for all i, j. As $\|\beta_M^\tau\|^2 = e+1$,

$$\beta_M^\tau = \sum_{\substack{0 \le i < q \\ 0 \le j < p}} (\pm \eta_{ij}) - \chi,$$

where $\|\chi\|^2 = 1$. Since $(\beta_M^\tau, (\overline{\psi} - \psi)^\tau) = 1$, either $\chi = \psi^{\tau_1}$ or $\chi = -\overline{\psi}^{\tau_1}$. $\quad\square$

(14.11.3) *Let* $G_0 = G - [\tilde{A}(M) \cup (W^\#)^G \cup (P^\#)^G \cup (Q^\#)^G]$. *If* $g \in G_0$, *then* $|\psi^{\tau_1}(g)| \ge 1$.

Proof. Since $g \notin \tilde{A}(M)$, $\beta_M^\tau(g) = 0$, and so, by (14.11.2), $\chi(g) = \sum_{i,j}(\pm\eta_{ij}(g))$ where $\chi = \psi^{\tau_1}$ or $-\overline{\psi}^{\tau_1}$. By (5.9), it suffices to show that $|\sum_{i,j}(\pm\eta_{ij}(g))| \ge 1$. Let a be an element of G of order divisible by p or by q. Then a is conjugate in G to an element of $C_G(x)$ for $x \in P^\# \cup Q^\#$. Suppose that $a \in C_G(x)$ for $x \in P^\#$ ($x \in Q^\#$, respectively). By (8.6.a), $a \in C_S(x)$ ($a \in C_T(x)$, respectively). If $a \in S - S'$ ($a \in T - T'$, respectively), then a is conjugate to an element of W by (2.1). If $a \in S'$ ($a \in T'$, respectively), then, by (14.4), (14.6) and (13.12), $a \in P^\#$ ($a \in Q^\#$, respectively). This proves that g has order prime to pq. By (3.9.c), we then have $\eta_{ij}(g) \in \mathbf{Z}$ for all i, j. By (3.9.a), for all i, j, there are indices r, s such that $\overline{\eta_{ij}} = \eta_{rs}$, and $\overline{\eta_{ij}} \ne \eta_{ij}$ for $(i,j) \ne (0,0)$. Since $\eta_{00}(g) = 1$, we see that $\sum_{i,j}(\pm\eta_{ij}(g)) \in 2\mathbf{Z} + 1$, and so $|\sum_{i,j}(\pm\eta_{ij}(g))| \ge 1$. $\quad\square$

(14.11.4) *Conclusion.*

Let ρ be the mapping defined in Hypothesis (7.1), with L and A of Hypothesis (7.1) replaced by M and $A(M)$. By (7.5),

$$\frac{1}{|G|}\left[\sum_{g \in G_0} |\psi^{\tau_1}(g)|^2 - |G_0| - |(W - (W_1 \cup W_2))^G| - |(P^\#)^G| - |(Q^\#)^G|\right]$$

$$+\|\psi^{\tau_1 \rho}\|^2 - \frac{|K^\#|}{|M|} \le 0.$$

By (14.11.3) and (7.8.b), it follows that

$$1 - \frac{pq}{k} \le \|\psi^{\tau_1 \rho}\|^2 \le 1 - \frac{1}{p} - \frac{1}{q} + \frac{1}{pq} + \frac{|P|-1}{|P|uq} + \frac{|Q|-1}{|Q|vp} + \frac{k-1}{kpq}.$$

Thus,

$$\frac{1}{p} + \frac{1}{q} \le \frac{pq}{k} + \frac{2}{pq} + \frac{1}{uq} + \frac{1}{vp}.$$

As UW_1 and VW_2 are Frobenius groups, $u > 2q$ and $v > 2p > 2q$. Thus,

$$\frac{2}{pq} + \frac{1}{uq} + \frac{1}{vp} < \frac{2}{q^2} + \frac{1}{2q^2} + \frac{1}{2q^2} = \frac{3}{q^2} \le \frac{1}{q}.$$

The preceding inequality implies that $\dfrac{1}{p} \le \dfrac{pq}{k}$. Now $v < \dfrac{k}{p} \le pq$ by (14.11.1). But, by (14.4), $v = 1+q+\cdots+q^{p-1} \equiv 1 \pmod{q}$, and $v \equiv 1 \pmod{p}$ because VW_2 is a Frobenius group. Thus $v > pq$, and we obtain a contradiction. $\square$

(14.12) *If L is conjugate to M in G, then Theorem (14.2) holds.*

Proof. If L is conjugate to M, then H is conjugate to K. By (14.11), (14.4) and (13.12), K is cyclic. Thus H is cyclic and, consequently, U is characteristic in H. By (14.7), Theorem (14.2) holds. $\square$

In (14.14) to (14.16), we continue to assume Hypothesis (14.3) and we will also assume

(14.13) Hypothesis. *L is not conjugate to M. Set $h = |H|$.*

(14.14) *One of the following two cases holds.*

(a) $(\beta_M^\tau, \varphi^{\tau_1}) \ne 0$ *and* $\dfrac{h-1}{pq} \le pq - 1.$

(b) $(\beta_L^\tau, \psi^{\tau_1}) \ne 0$, $q = 3$ *and* $p = 5$.

Proof. By (8.17.c), $\widetilde{A}_1(L) \cap \widetilde{A}_1(M) = \emptyset$. Thus, by (7.9), $(\beta_M^\tau, \varphi^{\tau_1}) \ne 0$ or $(\beta_L^\tau, \psi^{\tau_1}) \ne 0$. Moreover, it follows from (4.1) that $\mathcal{L}^{\tau_1}$ is orthogonal to $\mathcal{M}^{\tau_1}$. Suppose that $a = (\beta_M^\tau, \varphi^{\tau_1}) \ne 0$. Let $\mathcal{L} = \{\varphi_i \mid 1 \le i \le n\}$, where $\varphi_1 = \varphi$ and $\varphi_i(1) = a_i | L : H |$. Then $(\beta_M^\tau, (\varphi_i - a_i\varphi)^\tau) = 0$ for $2 \le i \le n$, and so $\beta_M^\tau = a \sum_{i=1}^{n} a_i \varphi_i^{\tau_1} + \Delta$ where Δ is orthogonal to $\mathcal{L}^{\tau_1}$. By (7.8.b), it follows that $\sum_{i=1}^{n} a_i^2 \le |M : K| - 1 = pq - 1$. Then (a) follows from the fact that

$$\sum_{i=1}^{n} a_i^2 = \frac{|L| - |L : H|}{|L : H|^2} = \frac{h-1}{pq}.$$

Again, if $(\beta_L^\tau, \psi^{\tau_1}) \ne 0$, we see that

$$\frac{v-1}{pq} = \frac{|K|-1}{|M : K|} \le |L : H| - 1 = pq - 1.$$

Suppose that $(\beta_L^\tau, \psi^{\tau_1}) \ne 0$. By (14.4), $v = (q^p - 1)/(q - 1)$. The inequality $(v-1)/(pq) < pq$ gives $q^p - q < (q-1)p^2q^2$, whence $q^p < p^2q^3$ and $q^{p-3} < p^2$. By (14.8.a), $q^{p-3} = q^{p+1}/q^4 > p^{q+1}/p^4 = p^{q-3}$. Thus $p^{q-3} < p^2$ and so $q = 3$. Let $f(x) = \dfrac{3^{x-3}}{x^2}$. Then $\dfrac{f(x+1)}{f(x)} = 3\left(1 - \dfrac{1}{x+1}\right)^2 > 1$ for $x \ge 2$. Thus, if $p \ge 7$, then $f(p) \ge f(7) = 3^4/7^2 > 1$. The inequality $q^{p-3} < p^2$ then implies that $p = 5$. $\square$

(14.15) $u = \dfrac{p^q - 1}{p - 1}.$

Proof. If this is not the case, then, by (14.6) and (13.15),

$$p \equiv 1 \pmod{q} \quad \text{and} \quad qu = \frac{p^q - 1}{p - 1}.$$

Let $h = ux$. By (14.5), $h \equiv 1 \pmod{p}$, whence $q \equiv qh = x(p^q - 1)/(p-1) \equiv x \pmod{p}$. There is thus an integer $n \in \mathbf{Z}$ such that $x = q + np$, and, since $q < p$, $n \geq 0$. As W_1 acts fixed-point-freely on U and on H, $x \equiv 1 \pmod{q}$ and so $np \equiv n \equiv 1 \pmod{q}$. Since x is odd, $x \geq q + (1+q)p$. Moreover, since $5 \not\equiv 1 \pmod{3}$, case (a) of (14.14) holds. As $x > pq$,

$$ h > \frac{p(p^q - 1)}{p - 1} > p^q + 1. $$

The inequality $(h-1)/(pq) \leq pq - 1$ thus implies that $p^q/(pq) < pq$, whence $p^{q-2} < q^2 < p^2$ and $q = 3$. But then $p < q^2 = 9$ and $p \equiv 1 \pmod{q}$, and so $p = 7$. But $x \geq q + (1+q)p = 31$ and $u = (7^2 + 7 + 1)/3 = 19$, and so $h \geq 31 \cdot 19$. The inequality $(h-1)/(pq) \leq pq - 1$ then gives $31 \cdot 19 - 1 \leq 20 \cdot 21$, which is a contradiction. $\qquad\qquad\square$

(14.16) $H = U$.

Proof. Suppose that $H \neq U$ and let $h = ux$. By (14.5) and (14.15),

$$ 1 \equiv h = ux \equiv x \pmod{p}. $$

Moreover, W_1 acts fixed-point-freely on U and on H, and so $x \equiv 1 \pmod{pq}$. Since x is odd and $x \neq 1$, $x > 2pq$. Suppose that case (a) of (14.14) holds. Then

$$ \frac{2pqu - 1}{pq} < \frac{h - 1}{pq} \leq pq - 1, $$

and so $2u < pq$. But $u \equiv 1 \pmod{p}$ and W_1 acts fixed-point-freely on U, whence $u \equiv 1 \pmod{pq}$, and we have a contradiction.

Now suppose that case (b) of (14.14) holds. Then

$$ u = (5^3 - 1)/4 = 31, h > 2pqu = 30 \cdot 31 \quad \text{and} \quad v = (3^5 - 1)/2 = 121. $$

Therefore,

$$ \frac{h - 1}{pq} \geq \frac{30 \cdot 31}{15} > \frac{120}{5} = \frac{v - 1}{p} > \frac{30}{3} = \frac{u - 1}{q}. $$

By (13.19.c) applied to S and to T, it follows that

$$ (\beta_L^\tau, \eta_{0j}) \equiv (\beta_L^\tau, \eta_{i0}) \equiv 1 \pmod{2} $$

for $i \geq 1$, $j \geq 1$. As β_L^τ vanishes on $W - (W_1 \cup W_2)$, we deduce as in (14.11.2) that $\beta_L^\tau = \sum_{i,j}(\pm\eta_{ij}) - \chi$, where $\chi = \varphi^{\tau_1}$ or $\chi = -\overline{\varphi}^{\tau_1}$. But this contradicts the fact that $(\beta_L^\tau, \psi^{\tau_1}) \neq 0$. $\qquad\qquad\square$

    Conclusion. By (14.12), (14.16) and (14.7), the proof of Theorem (14.2) is complete. As already remarked, Theorem (14.2) and [BG], Appendix C, show that G does not exist. $\qquad\qquad\square$

Notes

§ 2. This section corresponds to Section 9 of [FT], where an isometry is defined when A is 'tamely imbedded'. (Note that the set denoted by $\tilde{L}$ in [FT] is $A \cup \{1\}$.) The construction in [FT] was simplified and generalized by E. C. Dade in [D]. It is the version of Dade that we present here.

§ 3. Theorem (3.2) is Lemma 13.1 of [FT]. The original proof of this lemma used properties of the field generated by the character values. A combinatorial proof, analogous to that which is presented here, was given by M. J. Collins in [C].

§ 4. Hypothesis (4.6) is satisfied if L is a maximal subgroup of a minimal simple group of odd order which is of Type II, III, IV or V. The results of this section form part of Section 13 of [FT].

§ 5. Theorem (5.6) is a simplified version of Theorem 10.1 of [FT]. Hypothesis (5.2) replaces here the definition of a 'sub-coherent' subset of S given in [FT], Definition 10.2, which is more complicated, as well as the definition of a 'compatibly coherent' subset given in [Si2], which only applies when Hypothesis (4.6) holds.

§ 6. Theorem (6.3) and (6.5) are taken from Section 11 of [FT]. Theorem (6.8) is due to D. A. Sibley and represents an important simplification in comparison with the original proof of the Feit-Thompson Theorem. In the case in which L is a Frobenius group, this theorem was shown in [Si1], and also in [Is], Theorem 7.24. By using the theory of blocks, we can show that this theorem remains true if the hypothesis that $H^{\#}$ is a TI-subset is replaced by a more general hypothesis. On this topic, [Pu] and [Sm] can be consulted. Hypothesis (c2) of Theorem (6.8) is satisfied if L is a maximal subgroup of a minimal simple group of odd order which is of Type V.

§ 7. Theorem (7.11) was shown in [FHT]. Essentially the proof of [Su] is followed here; this is possible thanks to Sibley's Theorem (Theorem (6.8)). The lemmas used in this proof have been generalized in such a way that they can be used again in the subsequent material. In [FT], inequality (7.3) is only considered in the case in which it is an equality.

§ 8. Theorems (8.8) and (8.13) correspond to Theorems 14.1 and 14.2 of [FT]. An error slipped into the statement of Theorem 14.2 of [FT], where, in the notation of Theorem (8.13), it is stated that $A(M)$ is 'tamely imbedded'. According to the definition of a 'tamely imbedded subset' in [FT], if L is as in (8.13.c), the set $\bigcup_{x \in L_F^{\#}} C_L(x)^{\#} - L_F^{\#}$ should then be a TI-subset of G. But this property has not been established if L is not of Type I. This erroneous property is used in [FT], Lemma 10.3, to show that characters of G obtained by coherence satisfy the property of our (2.7), that of being constant on $aH(a)$ for $a \in A$. Theorem B of [BG], § 16, shows that certain properties which are shown to be valid only for Type I or only for Type II in Section 14 of [FT], are in fact valid for both of these types. Theorem E of [BG], § 16, is not explicit in Section 14 of [FT]. In property (8.13.c4), the type of L is due to Sibley and the fact that M is Frobenius was proved in 1991 by W. Feit and J. G. Thompson (cf. [BG], Corollary 15.9). Conditions (8.3.c) and (8.7.c) of the definitions of subgroups of Types I and V are stronger that those given in [FT].

§ 9. This section corresponds to Section 30 of [FT]. In [FT], the results are proven under the hypothesis that, if M is of Type III or IV, then S is not coherent. Complementary results on the coherence of S are proved in Section 30 of [FT], but are not needed here thanks to the method of Sibley for eliminating subgroups of Type V.

§ 10. This section corresponds to Sections 31 and 32 of [FT], where the maximal subgroups of Type V are eliminated. The proof is simplified by use of Sibley's Theorem (6.8). Also due to Sibley is the proof of (10.5) and the remark that Theorem (10.8) also applies to subgroups of Types III and IV.

§ 11. Results (11.3) to (11.7) come from Section 29 of [FT], where they are proven under the hypothesis that the set S of Hypothesis (10.1) is not coherent. A special case of (11.8) is proven in [FT], Lemma 33.4. The non-existence of maximal subgroups of Type IV is shown in a different way in [Si2].

§ 12. Theorem (12.7) is Theorem 33.1 of [FT]. Lemma (12.4) replaces [FT], Lemma 10.3, here. In Section 28 of [FT], a coherence result for certain maximal subgroups of Type I was shown. It is used in Section 33 of [FT] to show that, in the notation of (12.13), ψ is constant on $K^{\#}$. Lemma (12.5) replaces this result here.

§ 13. Results of Section 34 of [FT] are treated here. The conclusions of Lemmas 34.5, 34.6 and 34.7 of [FT] have been improved by G. Glauberman. These improvements are given here in (13.6), (13.9) and (13.11). They allow certain cases, given separate treatment in [FT], to be eliminated.

§ 14. This section corresponds to Sections 35 and 36 of [FT]. The order of the arguments has been rearranged to take account of the results of the preceding section and of (11.9).

References

[B] Bender, H. (1970) 'On the uniqueness theorem', *Illinois J. Math.* **14**, 376–384.

[BG] Bender, H.; Glauberman, G. (1994) *Local analysis for the odd order theorem*, London Mathematical Soc. Lecture Note Ser. **188**, Cambridge Univ. Press, Cambridge.

[C] Collins, M. J. (1988) 'Characters of finite groups having a self-normalising cyclic subgroup', *J. Algebra* **119**, 282–297.

[D] Dade, E. C. (1964) 'Lifting group characters', *Ann. of Math.* (2) **79**, 590–596.

[FHT] Feit, W.; Hall, M.; Thompson, J. G. (1960) 'Finite groups in which the centralizer of any non-identity element is nilpotent', *Math. Z.* **74**, 1–17.

[FT] Feit, W.; Thompson, J. G. (1963) 'Solvability of groups of odd order', *Pacific J. Math.* **13**, 775–1029.

[H] Huppert, B. (1967) *Endliche Gruppen*. I, Grundlehren Math. Wiss. **134**, Springer, Berlin.

[HB] Huppert, B.; Blackburn, N. (1982) *Finite groups*. III, Grundlehren Math. Wiss. **243**, Springer, Berlin.

[Is] Isaacs, I. M. (1976) *Character theory of finite groups*, Academic Press, New York.

[L] Lang, S. (1984) *Algebra*, Addison-Wesley, Reading, MA.

[Pe] Peterfalvi, T. (1984) 'Simplification du chapitre VI de l'article de Feit et Thompson sur les groupes d'ordre impair', *C. R. Acad. Sci. Paris Sér. I. Math.* **299**, 531–534.

[Pu] Puig, L. (1979) 'Structure locale et caractères', *J. Algebra* **56**, 24–42.

[Si1] Sibley, D. A. (1976) 'Coherence in finite groups containing a Frobenius section', *Illinois J. Math.* **20**, 434–442.

[Si2] Sibley, D. A. 'Lecture notes on character theory in the odd order paper', unpublished.

[Sm] Smith, S. D. (1977) 'Sylow automizers of odd order', *J. Algebra* **46**, 523–543.

[Su] Suzuki, M. (1957) 'The nonexistence of a certain type of simple groups of odd order', *Proc. Amer. Math. Soc.* **8**, 686–695.

Part II
A Theorem of Suzuki

Introduction

In this article, we present some simplifications of Suzuki's paper [S1].

We consider the following hypotheses.

(A1) *G is a finite group which acts doubly transitively on a set Ω, H is the stabilizer in G of a point in Ω, t is an involution in $G - H$, $D = H \cap H^t$ and Q is a subgroup of H such that $H = Q \rtimes D$. We assume that $|Q|$ is even and that $|D|$ is odd.*

(A2) *G acts faithfully on Ω.*

(A3) *G has 2-rank ≥ 2, i.e., it contains an elementary abelian subgroup of order 4.*

The theorem of Suzuki which we prove, states

Theorem A. *Under the above hypotheses, there is a normal subgroup L of G such that $|G/L|$ is odd and L is isomorphic to*

$$\mathrm{PSL}(2, q), \quad \mathrm{Sz}(q) \quad or \quad \mathrm{PSU}(3, q),$$

where q is a power of 2, $q > 2$, and where the action of L on Ω can be identified with the usual doubly transitive action of this group.

This theorem is used in the proof of the theorem of Bender [B] which classifies those finite groups G which have a proper subgroup H of even order such that $H \cap H^x$ is of odd order for $x \in G - H$. Bender's theorem has played an important role in the classification of finite simple groups. For its part, Suzuki's theorem is one of the components of the classification of those finite groups which have a split BN-pair of rank 1; this was accomplished in [HKS].

In the proof of the theorem, we use the following notation
$$K = \{x \in D \mid x^t = x^{-1}\}, \quad V = C_D(t) \quad \text{and} \quad W = C_V(K).$$
We assume as well the following induction hypothesis.

Every group of order $< |G|$ which satisfies the hypothesis of the theorem also satisfies its conclusion.

We take the classification of Zassenhaus groups as known (see [HB], Chapter XI). According to this classification, if a group G satisfying (A1) to (A3) is a Zassenhaus group, i.e., if the pointwise stabilizer in G of three points of Ω

is trivial, then G is isomorphic to $\mathrm{PSL}(2,q)$ or $\mathrm{Sz}(q)$. It is easy to see that G is a Zassenhaus group if and only if $V = 1$. Under the induction hypothesis, it then suffices to show:

If $V \neq 1$, either $O^{2'}(G) \neq G$ or $G \cong \mathrm{PSU}(3,q)$ where q is a power of 2.

The final characterization of G is effected by use of Suzuki's method of generators and relations which he used for the determination of the Zassenhaus groups. Every element x of $G - H$ can be expressed in a unique manner in the form $x = adtb$, where $a, b \in Q$ and $d \in D$. The structure of G is then determined by the structure of H, the action of t on D and the mappings $f, g : Q^{\#} \to Q^{\#}$ and $h : Q^{\#} \to D$ such that $txt = g(x)h(x)tf(x)$ for $x \in Q^{\#}$. These mappings satisfy certain relations which help to determine them.

But first we must determine the structure of H and eliminate certain cases in which theorems based on transfer guarantee that $O^{2'}(G) \neq G$.

To prove the theorem, we consider two cases. The first is that in which V has a subgroup P of prime order p such that $C_G(P)$ has 2-rank 1. The object is then to show that G has a quotient of order p. This case is dealt with in Chapter II. In it Wielandt's fixed point theorem is used to calculate $|Q|$; this simplifies the original proof in [S1].

In the second case, if P is a subgroup of V of prime order, then $C_G(P)$ has 2-rank ≥ 2; this makes application of the induction hypothesis possible.

The goal of Chapter III is the determination of H. The main obstacle is to show that Q is a 2-group. The proof of this fact is simplified here by the use of Sibley's Coherence Theorem [Si]. A proof of this theorem, in the form required here, is given in Appendix IV. Higman's classification of Suzuki 2-groups [Hi] then makes possible the determination of $Q \rtimes KW$.

In Chapter IV, we turn to the determination of the mappings f, g and h. If $V \neq W$, we must establish that $O^{2'}(G) \neq G$. In [S1], the proof of this fact contained an error which was noted and corrected in [KS]. In our version, this fact is a consequence of the determination of f.

This article reproduces the second part of [P] modified as required to make it independent of the first part of [P].

Notation

We denote inclusion in the broad sense by $\subset$ (e.g., $G \subset G$).

Unless stated otherwise, actions of a group on a set, or on a group, are right actions. If the group G acts on the set X and $a \in X$, then G_a denotes the stabilizer of a in G.

For q a power of 2, the field of q elements is denoted by $\mathbf{F}_q$. The group $PSL(2, q)$ is defined in [H], Kapitel II, §6. The group $PSU(3, q)$ is the group which is denoted $PSU(3, q^2)$ in [H], Kapitel II, §10. The *Suzuki group* $\mathrm{Sz}(q)$ is defined in [HB], Chapter XI, §3.

The notation $G = H \rtimes K$ means that H and K are subgroups of G with H normal in G, $G = HK$ and $H \cap K = 1$.

For a subset X of a group G, we write $X^\# = X - \{1\}$.

An element t of a group G is an *involution* if $t \neq 1$ and $t^2 = 1$.

Let G be a group and π a set of prime numbers. Then π' denotes the set of prime numbers which do not belong to π. The group G is a π-group if each of the prime divisors of $|G|$ is in π. We write $O_\pi(G)$ for the largest normal π-subgroup of G and $O^\pi(G)$ for the smallest normal subgroup of G such that $G/O^\pi(G)$ is a π-group. For $g \in G$, the π-*component* g_π and the π'-*component* $g_{\pi'}$ of g are those elements of $\langle g \rangle$ such that $g = g_\pi g_{\pi'}$ with $\langle g_\pi \rangle$ a π-group and $\langle g_{\pi'} \rangle$ a π'-group. If n is an integer, $n > 0$, then n_π is the largest integer such that n_π divides n and the prime divisors of n are in π. If $\pi = \{p\}$, we replace π by p in the above notations.

If G is a finite group, then $F(G)$ denotes the largest normal nilpotent subgroup of G.

If p is a prime number and P is a p-group, then $\Omega_1(P)$ denotes the subgroup of P generated by its elements of order p.

Let G be a finite group. Then $\mathrm{Irr}(G)$ denotes the set of irreducible complex-valued characters of G. If $\mathcal{X} \subset \mathrm{Irr}(G)$, then $\mathbf{Z}[\mathcal{X}]$ is the set of all $\mathbf{Z}$-linear combinations of elements of $\mathcal{X}$, and $\mathbf{Z}[\mathcal{X}]^\circ$ the set of $\alpha \in \mathbf{Z}[\mathcal{X}]$ such that $\alpha(1) = 0$. If $\alpha, \beta \in \mathbf{Z}[\mathrm{Irr}(G)]$, we write (α, β) for the usual scalar product of α and β. For a subgroup H of G, Res_H^G is *restriction*, $\mathbf{Z}[\mathrm{Irr}(G)] \to \mathbf{Z}[\mathrm{Irr}(H)]$, and Ind_H^G is *induction*, $\mathbf{Z}[\mathrm{Irr}(H)] \to \mathbf{Z}[\mathrm{Irr}(G)]$.

Chapter I. General Properties of G

§ 1. Consequences of Hypothesis (A1)

We assume in this section that G satisfies hypothesis (A1).

It follows from (A1) that the mapping which sends $\alpha \in \Omega$ to its stabilizer G_α in G is a bijection from Ω onto the set of conjugates of H in G. In what follows Ω is identified with $\{H^g \mid g \in G\}$ with G acting on this set by conjugation. For $X \subset G$, we will denote by Ω_X the set of fixed points of X on Ω. We will denote by I the set of involutions of G.

Throughout we will use the notation

$$K = \{x \in D \mid x^t = x^{-1}\}, \; V = C_D(t) \; \text{ and } \; W = C_V(K).$$

Proposition 1. (a) *Let $g \in G - H$. Then $H^g \cap H$ is conjugate to D in H. In particular, $|H^g \cap H|$ is odd.*

(b) *If $1 \neq X \subset Q$, then $N_G(X) \subset H$.*

(c) *Q contains a Sylow 2-subgroup of G.*

(d) *$N_G(Q) = N_G(H) = H$.*

(e) *If G has 2-rank ≥ 2, then $O_{2'}(G) = \bigcap_{x \in G} H^x$.*

Proof. (a) This follows immediately from the double transitivity of G.

(b) Let $g \in N_G(X)$. Then $X \subset H \cap H^g$. If $g \notin H$, then $Q \cap H \cap H^g = 1$ by (a), whence $X = 1$.

(c) By (A1), Q acts regularly on $\Omega - \{H\}$. Thus $|G| = |H| \cdot |\Omega| = |Q||D|(|Q| + 1)$, and $|D|(|Q| + 1)$ is odd.

(d) By (b), $N_G(Q) = H$ and, by (a), $N_G(H) = H$.

(e) By (a), $\bigcap_{x \in G} H^x$ is of odd order and so is contained in $O_{2'}(G)$. By (c), Q contains an elementary abelian subgroup X of order 4. But $O_{2'}(G)$ is generated by the subgroups $C_G(x) \cap O_{2'}(G)$ for $x \in X^{\#}$ ([HB], Chapter X, Lemma 1.9). Thus, by (b), $O_{2'}(G) \subset H$ whence $O_{2'}(G) \subset \bigcap_{x \in G} H^x$. $\quad\square$

Proposition 2. (a) *If $s \in H \cap I$ and $u \in I - (H \cap I)$, then su is of odd order.*

(b) *I is a conjugacy class of G.*

(c) *Suppose that $s \in Q \cap I$. Then the mapping $u \mapsto s^u$ is a permutation of the set $I - (H \cap I)$.*

(d) *The number of $u \in I$ such that $H^u = H^t$ is equal to $|H \cap I|$.*

Proof. (a) Suppose that su is of even order. Let w be the involution in $\langle su \rangle$. As s and u invert su, they centralize w. Since $s \in Q$, it then follows from Proposition 1(b) that $w \in Q$ and that $u \in Q$, which contradicts the hypothesis.

(b) The sets $H \cap I$ and $I - (H \cap I)$ are non-empty and, by (a), every element of $I - (H \cap I)$ is conjugate to every element of $H \cap I$.

(c) Let $s \in Q \cap I$. If $u \in I - (H \cap I)$, then $s^u \in I - (H \cap I)$ by Proposition 1(a). Let $v \in I - (H \cap I)$. By (a), sv is of odd order, whence s and v are conjugate in the dihedral group $\langle s, v \rangle$. It follows that there is an involution u in $\langle s, v \rangle$ for which $s^u = v$. As $v \notin H$, we see that $u \notin H$. Therefore the mapping $u \mapsto s^u$ from $I - (H \cap I)$ to $I - (H \cap I)$ is surjective and so bijective.

(d) By (c), the number of elements $u \in I$ for which $s^u \in H^t$ is equal to $|H^t \cap I| = |H \cap I|$. But $s^u \in H^t$ is equivalent to $H^u = H^t$ by Proposition 1(a). $\qquad\square$

Proposition 3. *We have $|K| = |H \cap I|$ and, if $s \in H \cap I$, then $s^K = H \cap I$.*

Proof. The mapping $k \mapsto kt$ is a bijection from K onto $Dt \cap I$. By Proposition 2(d), $|Dt \cap I| = |H \cap I|$ whence $|K| = |H \cap I|$. Let $s \in H \cap I$. If $x, y \in K$ satisfy $s^x = s^y$, then $s^{xt} = s^{yt}$ whence $x = y$ by Proposition 2(c). It follows that $|s^K| = |K| = |H \cap I|$. $\qquad\square$

Proposition 4. (a) *Every element of $G - H$ can be expressed in a unique manner in the form xty with $x \in H$ and $y \in Q$.*

(b) *There is a unique pair (s, r) such that $tst = r^{-1}tr$, $s \in H \cap I$ and $r \in Q$.*

(c) *Let $N = \bigcap_{x \in G} H^x$. Then $N = C_D(Q) \subset C_D(t)$. The group $\overline{G} = G/N$ acting on Ω satisfies (A1), $\overline{Q} \cong Q$ and, if s is as in (b), then the order of the image $\overline{st}$ of st in $\overline{G}$ is equal to the order of st.*

Proof. (a) As G is doubly transitive and t normalizes D, we see that
$$G - H = HtH = HtQ.$$
If $x_i \in H$, $y_i \in Q$ $(i = 1, 2)$ and $x_1 ty_1 = x_2 ty_2$, then
$$tx_2^{-1} x_1 t = y_2 y_1^{-1} \in H^t \cap Q = 1,$$
from which uniqueness follows.

(b) Let $u \in H \cap I$. By (a) and Proposition 3, it suffices to show that there is an element $k \in K$ such that $tu^k t$ is of the form $y^{-1} ty$ with $y \in Q$. By (a), there are elements $x \in H$ and $y \in Q$ such that $tut = xty$. It follows that $xtyxty = 1$ whence, with $a = yx$, $tat = a^{-1}$. Thus $a \in H \cap H^t$ and $a \in K$. For $k \in K$,
$$tk^{-1}ukt = ktutk^{-1} = ky^{-1}atyk^{-1} = ky^{-1}k^{-1} \cdot kak \cdot t \cdot kyk^{-1}.$$

Thus, $tu^k t$ is of the required form if and only if $kak = 1$, or $k^{-2} = a$. As $k \mapsto k^2$ is a permutation of K, there is indeed a unique element $k \in K$ such that $k^{-2} = a$.

(c) As Q is a normal subgroup of H acting regularly on $\Omega - \{H\}$, the stabilizer D of H^t in H acts in an equivalent manner on Q and on $\Omega - \{H\}$, whence $N = C_D(Q)$. As t is conjugate to an element of Q, t centralizes N. As $N \subset D$, $\overline{H} = \overline{Q} \rtimes \overline{D}$ in $\overline{G}$, $\overline{Q} \cong Q$ and $\overline{G}$ acting on Ω satisfies (A1). The elements of $\langle st \rangle \cap N$ are inverted by t, centralized by t and of odd order; it follows that $\langle st \rangle \cap N = 1$ and the order of $\overline{st}$ is the same as that of st. $\square$

In what follows, we set $\mathcal{N}(G) = \bigcap_{x \in G} H^x$. We will call the expression xty of Proposition 4(a) the *canonical form* of the element xty, the element s of Proposition 4(b) the *distinguished involution* of Q and the equation $tst = r^{-1}tr$ the *structure equation* of G. We note that, by the double transitivity of G and by the transitivity of D on $H \cap I$ (Proposition 3), G is transitive on the set of pairs (H', t') where H' is a conjugate of H and $t' \in I \cap (G - H)$. Thus the order of st is independent of the choice of $H \in \Omega$ and of $t \in I \cap (G - H)$.

In what follows, s is the distinguished involution of Q.

Lemma. *Let M be a finite group and let t be an element of order 2 in M. Let X be a subgroup of M of odd order normalized by t, $Y = C_X(t)$ and $Z = \{x \in X \mid x^t = x^{-1}\}$.*

(a) *The mappings $(y, z) \mapsto yz$ and $(y, z) \mapsto zy$ are bijections from $Y \times Z$ to X. In particular, $|X| = |Y||Z|$.*

(b) $\langle Z \rangle \lhd X$.

Proof. (a) Let $y_1, y_2 \in Y$ and $z_1, z_2 \in Z$ be such that $y_1 z_1 = y_2 z_2$. Then

$$z_1^2 = ((y_1 z_1)^t)^{-1}(y_1 z_1) = ((y_2 z_2)^t)^{-1}(y_2 z_2) = z_2^2,$$

whence $z_1 = z_2$ as $|X|$ is odd. The mapping $(y, z) \mapsto yz$ from $Y \times Z$ to X is therefore injective. Similarly, the mapping $(y, z) \mapsto zy$ is injective. It is thus sufficient to show that $|X| \leq |Y||Z|$. We know that $|t^X| = |X : Y|$ and that the mapping $u \mapsto tu$ from t^X to Z is injective and so $|X : Y| \leq |Z|$.

(b) We see that Y normalizes Z and that Z normalizes $\langle Z \rangle$ whence X normalizes $\langle Z \rangle$ by (a). $\square$

Proposition 5. $V = C_D(s)$ *and* $W = C_D(H \cap I)$.

Proof. If $v \in V$ and if (s, r) is as in Proposition 4(b), $ts^v t = (r^{-1})^v tr^v$. It follows from the uniqueness of (s, r) that $s^v = s$, whence $V \subset C_D(s)$. By part (a) of the lemma and Proposition 3, $|D : V| = |K| = |s^D| = |D : C_D(s)|$ so that $V = C_D(s)$. By Proposition 3, $k \mapsto s^k$ defines a bijection from K to $H \cap I$. As V normalizes K, it follows that $W = C_D(s) \cap C_D(K)$ is the set of elements of D which centralize $H \cap I$. $\square$

Proposition 6. *Let $X \subset D$ be such that $|\Omega_X| \geq 3$.*

(a) $C_G(X)$ *is doubly transitive on Ω_X and* $C_H(X) = C_Q(X) \rtimes C_D(X)$.

(b) $|C_Q(X)|$ *is even.*

(c) X *is conjugate in D to a subgroup of V.*

Proof. (a) By hypothesis Q acts regularly on $\Omega - \{H\}$. Let $H_1 \in \Omega_X - \{H\}$. There is an element $y \in Q$ such that $H^{ty} = H_1$. If $x \in X$, then

$$H^{ty} = H_1 = H_1^x = H^{tyx} = H^{tx^{-1}yx},$$

whence $y = x^{-1}yx$ and so $y \in C_Q(X)$. This shows that $C_Q(X)$ acts transitively, and so regularly, on $\Omega_X - \{H\}$ and that $C_H(X) = C_Q(X) \rtimes C_D(X)$. Similarly, $C_{Q^t}(X)$ acts transitively on $\Omega_X - \{H^t\}$ and so $C_G(X)$ acts doubly transitively on Ω_X since $|\Omega_X| \geq 3$.

(b) By (a), $|C_G(X)|$ is even. Let u be an involution in $C_G(X)$. Then there is an element $H' \in \Omega$ such that $u \in H'$. By Proposition 1(b), $X \subset C_G(u) \subset H'$ whence $H' \in \Omega_X$. As $C_G(X)$ is transitive on Ω_X, $C_H(X)$ is conjugate to $C_{H'}(X)$ in $C_G(X)$ whence $|C_H(X)| = |C_{H'}(X)|$ is even.

(c) There is an element $k \in K$ such that $s^k \in C_Q(X)$ by (b) and Proposition 3. Thus, $X \subset C_D(s^k) = V^k$ by Proposition 5. $\qquad\square$

§2. The Structure of Q and of K

We assume from this point onwards that G satisfies (A1), (A2) and (A3) and we retain the notation introduced in §1.

Proposition 1. (a) *If $x \in K - \{1\}$, then $C_Q(x) = 1$.*

(b) *Q is nilpotent.*

(c) *$H \cap I \subset Z(Q)$ and $(H \cap I) \cup \{1\}$ is an elementary abelian 2-group.*

Proof. (a) Let $x \in K$ and assume that $|\Omega_x| \geq 3$. By §1, Proposition 6(b), there is an element $u \in H \cap I$ such that $u^x = u$. But the mapping $y \mapsto u^y$ from K to $H \cap I$ is bijective (§1, Proposition 3) and so $x = 1$. Thus, if $x \in K - \{1\}$, $|\Omega_x| = 2$. As $C_G(x)$ acts on Ω_x, we see that $C_H(x) \subset D$ whence $C_Q(x) = 1$.

(b) By (A3) and §1, Proposition 3, $|K| = |H \cap I| > 1$. Thus there is an element $x \in K - \{1\}$. As $\langle x \rangle \subset K$, $\langle x \rangle$ acts without fixed points on Q by (a). Therefore Q is nilpotent by a theorem of Thompson ([H], Kapitel V, Hauptsatz 8.14).

(c) Since Q is nilpotent of even order, $Z(Q)$ contains an involution. As D acts transitively on $H \cap I$, it follows that $H \cap I \subset Z(Q)$ and that $(H \cap I) \cup \{1\}$ is an elementary abelian 2-group. $\qquad\square$

In what follows we will use the notation

$$Q_0 = (H \cap I) \cup \{1\} \quad \text{and} \quad q = |Q_0|,$$

and we will write $Q = S \times Q_1$ where S is the Sylow 2-subgroup of Q.

Proposition 2. *K is a cyclic normal subgroup of D.*

Proof. Let $\overline{D} = D/W$ and let A be such that $W \subset A \subset D$ and $\overline{A} = F(\overline{D})$. As $\overline{D}$ is faithful on Q_0 and transitive on $Q_0^{\#}$ (§1, Propositions 3 and 5), Proposition 1 of Appendix I shows that $\overline{A}$ acts fixed-point-freely on Q_0 and that $\overline{D}/\overline{A}$ is abelian.

By §1, Proposition 5, $\overline{A} \cap \overline{V} = 1$ whence $\overline{C_A(t)} = C_{\overline{A}}(t) = 1$. By §1, Lemma, $\overline{A}$ is contained in $J = \{\bar{a} \in \overline{D} \mid \bar{a}^t = \bar{a}^{-1}\}$. As D/A is abelian, the set B/A of those elements of D/A inverted by t is a group. Since t inverts the elements of B/A and of A/W, $C_B(t) \subset C_A(t) \subset W$ so that t inverts the elements of $\overline{B} = B/W$. Thus $\overline{B}$ is abelian and $\overline{B} = J$. By Fitting's Theorem ([H], Kapitel III, Satz 4.2), it follows that $\overline{A} = \overline{B} = J$.

As t acts on A, we have $A = (A \cap K)(A \cap V)$ (§1, Lemma). But since $\overline{A} = J$, it follows that $K \subset A$ and $A \cap V = W$ whence $A = KW$ and $|\overline{A}| = |K|$.

By Appendix I, $\overline{A}$ is cyclic and so there is an element $k \in K$ such that $\overline{A} = \langle \overline{k} \rangle$. Then $|\langle k \rangle| \geq |\langle \overline{k} \rangle| = |\overline{A}| = |K|$, whence $K = \langle k \rangle$. Finally, K is normal in D by §1, Lemma (b).   □

Corollary. *The Sylow 2-subgroup S of Q is either abelian or a Suzuki 2-group.*

(See Appendix III, Definition 1.)

Proof. Indeed, K acts regularly on $S \cap I = H \cap I$ (§1, Proposition 3).   □

If F is a field and A a subgroup of $\mathrm{Aut}(F)$, we put $\mathcal{L}(F, A) = (F \rtimes F^*) \rtimes A$ where F^* acts on the additive group F by multiplication on the right and A acts naturally on F and on F^*.

Proposition 3. *There is a group A of automorphisms of $\mathbf{F}_q$ and an isomorphism from $Q_0 \rtimes (D/W)$ to $\mathcal{L}(\mathbf{F}_q, A)$ which identifies Q_0 with the additive group $\mathbf{F}_q$, K with $\mathbf{F}_q^*$ and V/W with A. In particular, V/W is cyclic.*

Proof. As $\overline{K} = KW/W$ is a cyclic normal subgroup of $\overline{D} = D/W$ which acts transitively on $Q_0^{\#}$, Proposition 2 of Appendix I applies. By (a) of that proposition, Q_0 is a vector space of dimension 1 over $\mathbf{F}_q$ and K is identified with the group of multiplications of this space by non-zero scalars. We identify the $\mathbf{F}_q$-space Q_0 with $\mathbf{F}_q$ in such a way that the distinguished involution s of Q is identified with $1 \in \mathbf{F}_q$. By §1, Proposition 5 and by Appendix I, Proposition 2(b), $\overline{V} = V/W$ acts on $Q_0 \rtimes K \cong \mathbf{F}_q \rtimes \mathbf{F}_q^*$ as a group of field automorphisms.   □

§3. Application of the Induction Hypothesis

Lemma 1. *Assume that G satisfies the conclusion of Theorem A and let L be the subgroup of G given by the theorem. Then Q is a 2-group and*

$$L = O^{2'}(G) = \langle Q^x \mid x \in G \rangle.$$

Proof. From the structure of the permutation groups $\mathrm{PSL}(2, q)$, $\mathrm{Sz}(q)$ and $\mathrm{PSU}(3, q)$, we know that $|\Omega| - 1$ is a power of 2 (see [HB], Chapter XI, Example 1.3 a) and Theorem 3.3, and [H], Kapitel II, Satz 10.12). As Q acts regularly on $\Omega - \{H\}$, Q is a 2-group. It follows that $Q \subset L$ and that Q is a Sylow 2-subgroup of G (§1, Proposition 1(c)). Thus, $O^{2'}(G) = \langle Q^x \mid x \in G \rangle \subset L$. But we know that L is simple ([H], Kapitel II, Hauptsatz 6.13, [HB], Chapter XI, Theorem 3.6, and [H], Kapitel II, Satz 10.13). Therefore, $O^{2'}(G) = L$. $\square$

We now assume that every group of order $< |G|$ which satisfies the hypothesis of Theorem A also satisfies its conclusion.

Proposition 1. *Let X be a subgroup of V, $X \neq 1$.*

(a) *The group $L = C_G(X)$ acting on Ω_X satisfies* (A1); *moreover,*

$$\mathcal{N}(L) = C_{L \cap D}(L \cap Q) \subset L \cap V.$$

(b) $N_G(X) = C_G(X)N_V(X)$.

(c) *Assume that $C_G(X)$ has 2-rank ≥ 2. Let $F = O^{2'}(C_G(X))$ and let $\ell = |C_{Q_0}(X)|$. Then $C_{Q_1}(X) = 1$, $\mathcal{N}(L) \cap F = Z(F)$ and one of the following three cases holds.*

(i) $F/Z(F) \cong \mathrm{PSL}(2, \ell)$, *st has order 3 and $C_Q(X)$ is elementary abelian of order ℓ.*

(ii) $F/Z(F) \cong \mathrm{Sz}(\ell)$, *st has order 5 and $C_Q(X)$ is a Suzuki 2-group of type A and order ℓ^2.*

(iii) $F/Z(F) \cong \mathrm{PSU}(3, \ell)$, *st has order 3 and $C_Q(X)$ is a Suzuki 2-group of order ℓ^3.*

Proof. (a) By §1, Proposition 5, the three elements H, H^t and H^{ts} of Ω are fixed by X. Therefore L satisfies (A1) by §1, Proposition 6. The statement concerning $\mathcal{N}(L)$ has been seen in §1, Proposition 4(c).

(b) Let $g \in N_G(X)$. As $C_G(X)$ is doubly transitive on Ω_X by (a), there is an element $f \in C_G(X)$ such that gf fixes H and H^t. It follows that $N_G(X) = C_G(X)N_D(X)$. By §1, Lemma, $N_D(X) = N_K(X)N_V(X)$ and, since $K \lhd D$, $[N_K(X), X] \subset X \cap K = 1$ whence $N_D(X) \subset C_G(X)N_V(X)$.

(c) By (a) and the induction hypothesis, $L/\mathcal{N}(L)$ satisfies the conclusion of Theorem A. By Lemma 1, $C_Q(X)$ is a 2-group and so $C_{Q_1}(X) = 1$. By (a) $\mathcal{N}(L)$ centralizes $C_Q(X)$ and so centralizes $F = \langle C_Q(X)^x \mid x \in L \rangle$. As $Z(F)$ centralizes $C_Q(X)$ and $C_{Q^t}(X)$, we see that $Z(F) \subset D$ (§1, Proposition 1(b)) so that $Z(F) \subset \mathcal{N}(L) = C_{L \cap D}(L \cap Q)$ and $\mathcal{N}(L) \cap F = Z(F)$. We know that $F/Z(F) = O^{2'}(L/\mathcal{N}(L))$ so that, by the induction hypothesis and Lemma 1, $F/Z(F)$ is isomorphic to one of the groups listed for some power ℓ of 2. By §1, Proposition 4(b), the structure equation of L is the same as that of G and so $\overline{s}$ is the distinguished involution of $\overline{L} = L/\mathcal{N}(L)$. By §1, Proposition 4(c), the order of st is equal to the order of $\overline{st}$ in $\overline{L}$ and $C_Q(X) \cong \overline{C_Q(X)}$ is the Sylow 2-subgroup of $\overline{C_H(X)}$. The identity $\ell = |C_{Q_0}(X)|$ and the statements concerning the order of st and the structure of $C_Q(X)$ are therefore consequences of the structures of the groups $\mathrm{PSL}(2, \ell)$, $\mathrm{Sz}(\ell)$ and $\mathrm{PSU}(3, \ell)$.

For the Sylow 2-subgroups of $\mathrm{PSL}(2,\ell)$ and $\mathrm{Sz}(\ell)$, see [H], Kapitel II, Satz 8.2 and [HB], Chapter XI, Lemma 3.1 and Theorem 3.3. The structure of the Sylow 2-subgroup of $\mathrm{PSU}(3,\ell)$ is given in [H], Kapitel II, Satz 10.12. For the order of st in $\mathrm{PSL}(2,\ell)$ and $\mathrm{Sz}(\ell)$, where s is the distinguished involution, see [HB], Chapter XI, Examples 10.7 a) and 10.7 b). Lastly, using the presentation of $\mathrm{PSU}(3,\ell)$ given in [H], Kapitel II, Satz 10.12, with s and t the images in $\mathrm{PSU}(3,\ell)$ of the matrices

$$\begin{pmatrix} 1 & 0 & 1 \\ 0 & 1 & 0 \\ 0 & 0 & 1 \end{pmatrix} \quad \text{and} \quad \begin{pmatrix} 0 & 0 & 1 \\ 0 & 1 & 0 \\ 1 & 0 & 0 \end{pmatrix}$$

respectively, we see that $s^2 = t^2 = 1$ and $tst = sts$, whence s is the distinguished involution of Q relative to t, and st is of order 3. $\qquad\square$

Proposition 2. *If G is not simple, then the conclusion of Theorem A holds for G.*

Proof. Let L be a normal subgroup of G with $1 \neq L \neq G$. By (A2), (A3) and §1, Proposition 1(e), $O_{2'}(G) = 1$. Thus, $I \subset L$ and we may assume that $L = \langle I \rangle$. By §1, Proposition 3, there is an element $k \in K^\#$. Then $k = t(tk) \in L$ since t and $tk \in I$. Let $x, y \in Q$. If $x^{-1}x^k = y^{-1}y^k$, then $yx^{-1} = (yx^{-1})^k$ and so $x = y$ by §2, Proposition 1(a). The mapping $x \mapsto [x, k]$ from Q to itself is then injective and so bijective. Thus, $Q = [Q, k] \subset L$. As Q acts regularly on $\Omega - \{H\}$ and as $t \in L$, we see that L acting on Ω satisfies (A1), (A2) and (A3). Since L acts doubly transitively on Ω, $G = LD$ and $|G : L|$ is odd. The conclusion then follows from the induction hypothesis applied to L. $\qquad\square$

We conclude with some technical lemmas which will be needed later.

Lemma 2. *Let X and Y be subsets of V which are conjugate in G. Then X and Y are conjugate in V.*

Proof. Let $g \in G$ be such that $Y = X^g$. Then H, H^t, H^g and H^{tg} belong to Ω_Y. By Proposition 1(a), $C_G(Y)$ is doubly transitive on Ω_Y, and so there is an element $h \in C_G(Y)$ such that $H^{gh} = H$ and $H^{tgh} = H^t$. Then $X^{gh} = Y$ and $gh \in D$. On applying the canonical homomorphism from D to V to the two sides of $X^{gh} = Y$, we see that X and Y are conjugate in V. $\qquad\square$

Recall that an element of G is called *strongly real* if it is the product of two involutions.

Lemma 3. *Let x be a strongly real element of G such that $x^2 \neq 1$. Then x is conjugate in G to an element of the form ut with $u \in Q_0^\#$, and $|C_G(x)|$ is odd.*

Proof. Let $x = uv$ with $u, v \in I$. As $x^2 \neq 1$, the conjugates of H which contain u and v are distinct by §2, Proposition 1(c). By the double transitivity of

G and by §1, Proposition 3, G is transitive on the set of triples (H_1, H_2, v) where H_1, $H_2 \in \Omega$, $H_1 \neq H_2$ and $v \in H_2 \cap I$. Thus x is conjugate to an element of the form ut with $u \in Q_0^\#$.

Let $u \in Q_0^\#$ and $x = ut$. Let $y \in I \cap N_G(\langle x \rangle)$. If $y \notin H$, then uy is of odd order (§1, Proposition 2(a)) whence y is conjugate to u in $\langle y, u \rangle \subset N_G(\langle x \rangle)$, so that $y \notin C_G(x)$. Similarly, if $y \in H$, y is conjugate to t in $\langle y, t \rangle$ so that $y \notin C_G(x)$. Thus $|C_G(x)|$ is odd. $\qquad\square$

Lemma 4. *Suppose that st is of order 3 and that $V \neq 1$. Then $\langle Q_0, K, t \rangle = Q_0 K \cup Q_0 K t Q_0$ is isomorphic to $\mathrm{PSL}(2, q)$.*

Proof. Since st is of order 3, $tst = sts$. Thus, for $k \in K$,
$$ts^k t = ktstk^{-1} = ksk^{-1} \cdot k^2 t \cdot ksk^{-1},$$
whence $tQ_0 t \subset Q_0 K t Q_0$. Taking account of the fact that t normalizes K and K normalizes Q_0, we see that $Q_0 K \cup Q_0 K t Q_0$ is a subgroup of G, whose order is $q(q-1)(q+1)$. Since $V \neq 1$, this subgroup has order $< |G|$ and satisfies the hypothesis of Theorem A with Q_0 and K in place of Q and D. By the induction hypothesis, this subgroup is isomorphic to $\mathrm{PSL}(2, q)$. $\qquad\square$

Lemma 5. *Suppose that st is of order 3 and that Q is a Suzuki 2-group of order q^3. Then W is a cyclic group and $|W|$ divides $q + 1$. If, further, $W \neq 1$, then Q is a Suzuki 2-group of type B.*

(See Appendix III, Definition 3.)

Proof. Let $w \in W^\#$. As $Q_0 \subset C_Q(w)$ and st is of order 3, Proposition 1 shows that $C_Q(w) = Q_0$ or Q; but, as D acts faithfully on $\Omega - \{H\}$, we have $C_Q(w) = Q_0$. Let X be a K-subgroup of Q of order q^2 (such a subgroup exists by the theorem of Appendix III). Suppose that $X^w = X$. As w centralizes s and $|\{x \in X \mid x^2 = s\}| = (q^2 - q)/(q - 1) = q$, w centralizes an element of $\{x \in X \mid x^2 = s\}$; this contradicts the fact that $C_Q(w) = Q_0$. Consequently, $X^w \neq X$ and, since w centralizes K, X^w is a K-subgroup of Q isomorphic to X as a K-group. By the theorem of Appendix III, it follows that, if $W \neq 1$, then Q is a Suzuki 2-group of type B and that K can be identified with $\mathbf{F}_q^*$ and $\overline{Q} = Q/Q_0$ with a 2-dimensional vector space over $\mathbf{F}_q$ in such a way that K acts on $\overline{Q}$ as the group of multiplications of $\overline{Q}$ by non-zero scalars.

Thus W is identified with a subgroup of $GL(\overline{Q}) \cong GL(2, q)$. Moreover, W acts without fixed points on the set of $q + 1$ K-subgroups of order q^2 in Q, whence $|W|$ divides $q + 1$, and W is cyclic by the structure of the subgroups of $GL(2, q)$ (see [H], Kapitel II, §8). $\qquad\square$

108

Chapter II. The First Case

For the proof of the main theorem of this chapter, the Hall-Wielandt Theorem (see [Ha], Theorem 14.4.2) is needed. Let G be a finite group, p a prime number and P a Sylow p-subgroup of G. A subgroup A of P is said to be *weakly closed in P relative to G* if, for all $g \in G$, $A^g \subset P$ implies that $A^g = A$. Let $Z_{p-1}(P)$ be the $(p-1)$st term of the upper central series of P. Let A be a weakly closed subgroup of P relative to G. The Hall-Wielandt Theorem states that, if $A \subset Z_{p-1}(P)$, or if $p > 2$ and A is abelian, then $G/O^p(G)$ is isomorphic to $N_G(A)/O^p(N_G(A))$.

In this chapter it will be assumed that

(B1) *The subgroup V contains a subgroup P of prime order p such that $C_G(P)$ has 2-rank 1.*

We will demonstrate the following theorem.

Theorem B. *The conclusion of Theorem A holds for G under the hypothesis* (B1).

If G has a normal subgroup of index p, the conclusion of the theorem holds by Chapter I, §3, Proposition 2. **It will be assumed therefore that**

(B2) *G has no normal subgroup of index p.*

(1) $V = W \rtimes P$, $|Q_0| = 2^p$, $N_G(P) = C_G(P)$ and $C_D(P) = C_W(P) \times P$.

Proof. As $P \cap W = 1$, P acts as a group of automorphisms on the field $Q_0 \cong \mathbf{F}_q$ and on K as on $\mathbf{F}_q^*$ (Chapter I, §2, Proposition 3). Since $|C_{Q_0}(P)| = 2$ by (B1), it follows that $|Q_0| = 2^p$ and that $C_K(P) = 1$. Since V/W can be identified with a subgroup of $\mathrm{Aut}(\mathbf{F}_{2^p})$, $V = W \rtimes P$. Moreover, $N_G(P) = C_G(P)N_V(P)$ (Chapter I, §3, Proposition 1(b)); but $N_V(P) = C_W(P)P$ and so $N_G(P) = C_G(P)$. Furthermore, $C_D(P) = C_K(P)C_V(P) = C_V(P) = C_W(P)P$. $\square$

(2)(a) *The group $C_G(P)$, in its action on Ω_P, satisfies* (A1); *its kernel is* $N = C_D(C_Q(P)) \cap C_G(P)$.

(b) *There is a near-field F such that $C_G(P)/N = (F \rtimes C_Q(P)) \rtimes \Sigma$ with $C_Q(P) \cong F^*$, with $\Sigma = \overline{C_W(P)}$ identified with a group of automorphisms of F and with $C_Q(P) \rtimes \Sigma$ acting on F in its natural manner.*

Proof. Part (a) follows from Chapter I, §3, Proposition 1(a); part (b) then follows from Chapter I, §1, Proposition 4(c) and from Proposition 1 of Appendix II. $\square$

(3) *For each prime divisor r of $|Q_1|$, there is an integer i, $0 \le i \le p-1$, such that $r \equiv 2^i \pmod{2^p - 1}$. In particular, $r \ne p$ so that $(|Q|, |KP|) = 1$.*

Proof. Let M be an elementary abelian r-subgroup of Q_1 normalized by KP such that $M \neq 1$ and M is minimal with these properties. We denote the law of composition of M additively. As KP is a Frobenius group and K acts fixed-point-freely on M, we see that $C_M(P) \neq 0$ and, by (2)(b), $C_M(P)$ is of dimension 1 over $\mathbf{F}_r$. (We use the fact that, in a Frobenius complement, every subgroup whose order is the product of two primes is cyclic: [H], Kapitel V, Satz 8.15.)

Moreover, from [Is], Theorem 15.16, we see that $\dim M = p \dim C_M(P) = p$ (dimensions are over $\mathbf{F}_r$). By Clifford's Theorem ([Is], Theorem 6.5), M is the direct sum of irreducible $\mathbf{F}_r[K]$-modules, all having the same dimension, and, as $\dim M$ is prime, either M has an $\mathbf{F}_r[K]$-submodule of dimension 1 over $\mathbf{F}_r$ or M is irreducible as an $\mathbf{F}_r[K]$-module.

In the first case, as K acts fixed-point-freely on M and $|K| = 2^p - 1$, we obtain $r \equiv 1 \pmod{2^p - 1}$. Assume that the second case holds. By Proposition 2 of Appendix I, we can endow M with the structure of a field such that $K \rtimes P$ is a group of semilinear transformations $M \to M$, and, as $\dim M = p$, $P = \mathrm{Aut}(M)$. Therefore there is an element $a \in P^{\#}$ such that $x^a = x^r$ for all $x \in K$. By Chapter I, §2, Proposition 3, $K \rtimes P$ can also be identified with the group of non-zero semilinear transformations $\mathbf{F}_{2^p} \to \mathbf{F}_{2^p}$. Thus there is an integer i such that $0 \leq i \leq p - 1$ and $x^a = x^{2^i}$ for all $x \in K$. It follows that $r \equiv 2^i \pmod{2^p - 1}$. $\qquad\square$

(4) $|Q| = |C_Q(P)|^p = |F^*|^p$.

Proof. If X is a subset of KP, we denote by $\underline{X}$ the sum of the elements of X taken in the algebra $\mathbf{Z}[KP]$. Since KP is a Frobenius group with kernel K, in $\mathbf{Z}[KP]$ we have

$$\sum_{x \in K} \underline{P^x} + \underline{K} = \underline{KP} + |K| \cdot \underline{1}.$$

Applying Wielandt's fixed point theorem ([HB], Chapter XI, Theorem 12.4) to the action of KP on Q, we then obtain $|C_Q(P)|^p = |Q|$. $\qquad\square$

(5) *Assume that F is not a field, that is, that $C_Q(P)$ is not abelian. Then F is isomorphic to the near-field $F_{9,2}$ and $Q_1 = 1$.*

Proof. By (2), $C_Q(P)$ is a Frobenius complement. As Q is nilpotent, it follows that $C_{Q_1}(P)$ is cyclic, and, if $C_Q(P)$ is not abelian, $C_S(P)$ is not abelian. By Chapter I, §2, Corollary to Proposition 2, S is then a Suzuki 2-group and so of exponent 4. As $C_S(P)$ is of 2-rank 1, it is thus quaternion of order 8 ([H], Kapitel III, Satz 8.2).

The group $F^* \cong C_Q(P)$ has a cyclic subgroup of index 2 and so F is isomorphic to $F_{r^2,2}$ and $|Z(F^*)| = r - 1$ (Appendix II, Proposition 2). From the structure of $C_Q(P)$, we obtain $|F^*/Z(F^*)| = 4$; but
$$|F^*/Z(F^*)| = (r^2 - 1)/(r - 1) = r + 1,$$
and so $r = 3$, $|C_Q(P)| = 8$ and $Q_1 = 1$ by (4). $\qquad\square$

Let f be the order of st. Thus f is the characteristic of F by (2), Chapter I, §1, Proposition 4(c) and Appendix II, Proposition 1.

(6) *Assume that $Q_1 = 1$. If $F \cong F_{9,2}$, then $|\Sigma| = 1$ or 3. Otherwise, $|F| = f$ or 9, and $\Sigma = 1$.*

Proof. An odd order group of automorphisms of $F_{9,2}$ can only have order 1 or 3 as $F_{9,2}^*$ is quaternion of order 8. Suppose that $F \not\cong F_{9,2}$. By (5), F is a field while F^* is a 2-group by hypothesis. If $|F| = f^a$, there is an integer b for which $f^a = 2^b + 1$. By a lemma from arithmetic ([HB], Chapter IX, Lemma 2.7), it follows that $a = 1$ or $f^a = 9$. As $|\Sigma|$ is odd, we obtain $\Sigma = 1$. $\qquad\square$

(7) $N = P$ *and* $\Sigma \cong C_W(P)$.

(N was defined in (2)(a).)

Proof. By (1), $N = (N \cap W) \times P$. Suppose that $N \cap W \neq 1$. If $R = C_Q(N \cap W)$, then, by Chapter I, §3, Proposition 1(c), either $f = 3$ and $|R| = |Q_0|$ or $|Q_0|^3$, or $f = 5$ and $|R| = |Q_0|^2$. As N centralizes $C_Q(P)$, we have $C_Q(P) \subset R$, and Q is a 2-group by (4). Then, by (6) and (2)(b), one of the following three cases holds.

    (a) $|F| = 3$, $|C_Q(P)| = 2$.

    (b) $|F| = 9$, $|C_Q(P)| = 8$.

    (c) $|F| = 5$, $|C_Q(P)| = 4$.

By (4), we have for each case respectively.

    (a) $|Q| = |Q_0|$.

    (b) $|Q| = |Q_0|^3$.

    (c) $|Q| = |Q_0|^2$.

As $N \cap W$ acts faithfully on Q, $Q_0 \subset R \subsetneq Q$ so that case (a) is impossible. In cases (b) and (c), $C_Q(P)$ has an element of order 4 since $|C_{Q_0}(P)| = 2$, whence $Q_0 \subsetneq R \subsetneq Q$. Thus (c) is impossible while, in case (b), $|R| = |Q_0|^2$ whence $f = 5$, a contradiction. $\qquad\square$

(8) *Assume that $Q_1 \neq 1$. Let $\ell = |\Sigma|$. If $\ell \neq 1$, then ℓ is prime and F is a field of order 3^ℓ, 5^ℓ or 9^ℓ.*

Proof. By (5), F is a field. Let $w \in C_W(P)^\#$. It follows from the induction hypothesis that $f = 3$ or 5 and that $C_Q(w)$ is a 2-group so that $C_{F^*}(w)$ is a 2-group. Since $C_{F^*}(w)$ is the multiplicative group of the field of fixed points of w in F, there are integers a and b such that $|C_F(w)| = f^a$ and $f^a = 2^b + 1$. Therefore $|C_F(w)| = f$ or 9. Moreover, if $f = 3$, there cannot be elements w_1 and w_2 in $C_W(P)^\#$ such that $|C_F(w_i)| = 3^i$ $(i = 1, 2)$ as otherwise w_1 would be of even order. Thus $|C_F(w)|$ is independent of $w \in C_W(P)^\#$ and so $\ell = |C_W(P)|$ is prime and $|F| = |C_F(w)|^\ell$. $\qquad\square$

(9) $p = f$.

Proof. Let T be the transfer from G to $H/(QKW)$ ([H], Kapitel IV, §1). Let $x \in P^{\#}$. By (B2), $T(x) = 1$. By Chapter I, §1, Proposition 4(a), the identity element 1 and the elements ty for $y \in Q$ form a system of representatives for the right cosets of H in G. For $y \in Q$, $tyx = xty^x$. By the definition of T, we then obtain $T(x) = x^{|Q|+1}(QKW)$. As $P \cap (QKW) = 1$, we have $x^{|Q|+1} = 1$ and so p divides $|Q| + 1$. By (4),

$$|F| = |F| - 1 + 1 \equiv (|F| - 1)^p + 1 = |Q| + 1 \equiv 0 \pmod{p}.$$

It follows that $p = f$. $\qquad\qquad\qquad\qquad\qquad\qquad\qquad\qquad\qquad\qquad\qquad\square$

(10) *Let $|F| = p^m$. One of the following two cases holds.*

    (10.1) $p \nmid |\Sigma|$ *and* $|G|_p = p^{m+2}$.

    (10.2) $p = |\Sigma| = 3$, $F \cong F_{9,2}$, W *is cyclic of order* 3 *or* 9 *and* $|G|_3 = 3^4|W|$.

Proof. By (4),

$$
\begin{aligned}
|Q| + 1 &= 1 + (-1 + p^m)^p \\
&= 1 - 1 + p \cdot p^m - \binom{p}{2} p^{2m} + \cdots + p^{pm} \\
&\equiv p^{m+1} \pmod{p^{m+2}},
\end{aligned}
$$

and so $(|Q|+1)_p = p^{m+1}$. Thus $|G|_p = (|Q|+1)_p |H|_p = p^{m+2}|W|_p$. If $p \nmid |\Sigma|$, then $p \nmid |C_W(P)|$ and so $p \nmid |W|$ whence $|G|_p = p^{m+2}$. Suppose that $p \mid |\Sigma|$. If $Q_1 \neq 1$, then, by (8), F is a field of order 3^3, 5^5 or 9^3 and

$$|C_Q(P)| = |F^*| = 2 \cdot 13,\ \ 4 \cdot 11 \cdot 71\ \ \text{or}\ \ 8 \cdot 7 \cdot 13$$

respectively. But this contradicts (3) since the cases $13 \equiv 2^i \pmod{7}$ for $0 \le i \le 2$ and $11 \equiv 2^i \pmod{31}$ for $0 \le i \le 4$ are impossible. Thus $Q_1 = 1$ and, by (6), $p = |\Sigma| = 3$ and $F \cong F_{9,2}$. In this case, $|Q| = |F^*|^p = 8^3$ and $C_Q(P) \cong F^*$ is not abelian, whence Q is a Suzuki 2-group. Then W is cyclic of order 3 or 9 by Chapter I, §3, Lemma 5. Furthermore, $|G|_3 = 3^{m+2}|W|_3 = 3^4|W|$. $\qquad\square$

Remark. If (B2) is not assumed, then, for G isomorphic to $\mathrm{PSU}(3,8) \rtimes \mathrm{Aut}\,\mathbf{F}_8$ or to $\mathrm{PGU}(3,8) \rtimes \mathrm{Aut}\,\mathbf{F}_8$, it is easy to see that case (10.2) holds.

(11) *Let R be the inverse image of F in G. Then $R = T \times P$, where T is a subgroup normalized by $C_Q(P)C_W(P)$, and $T \rtimes C_Q(P) \cong F \rtimes F^*$. Furthermore, $C_Q(P)$ acts regularly on $\mathcal{A} - \{P\}$, where $\mathcal{A}$ denotes the set of subgroups of R of order p which are not contained in T.*

Proof. Assume that R is not abelian. Then $[R, R] = Z(R) = P$ since $C_Q(P)$ acts transitively on F^*. Also $N_G(R) \subset N_G(P)$, and, in case (10.1), R is a Sylow p-subgroup of $N_G(P)$ and so of G. But $|R| = p^{m+1}$, contrary to (10).

In case (10.2), $RC_W(P)$ is a Sylow 3-subgroup of $N_G(P)$. As Σ acts non-trivially on $R/P \cong F$, it follows that $Z(RC_W(P)) = Z(R) = P$ whence $N_G(RC_W(P)) \subset N_G(P)$ and $RC_W(P)$ is a Sylow 3-subgroup of G. But $|RC_W(P)| = 3^4$, contrary to (10).

Thus R is abelian and the first assertion has been verified with $T = [R, s]$.

Now $|\mathcal{A}| = (p^{m+1} - p^m)/(p-1) = p^m = |F|$. For the second assertion, it is then sufficient to show that an element a of $C_Q(P)^{\#}$ normalizes no element of $\mathcal{A} - \{P\}$. Let $P_1 \in \mathcal{A}$ be such that a normalizes P_1. Since a normalizes T and centralizes $R/T \cong P$, $[a, P_1] \subset P_1 \cap T = 1$ and a centralizes P_1. But a acts fixed-point-freely on R/P so that $C_R(a) = P$, whence $P_1 = P$. $\qquad\square$

(12) *Case* (10.2) *holds.*

Proof. Assume that case (10.1) holds.

By (10), R is not a Sylow p-subgroup of $N_G(R)$ so that $N_G(P) \subsetneq N_G(R)$. Let $\mathcal{A}'$ be the orbit of P under the action of $N_G(R)$. Let $P_1 \in \mathcal{A}'$. As P_1 is conjugate to P, the elements of $P_1^{\#}$ are not strongly real (Chapter I, §3, Lemma 3). The elements of T are inverted by s and so $P_1 \cap T = 1$ and $\{P\} \subsetneq \mathcal{A}' \subset \mathcal{A}$. Since $C_Q(P)$ acts regularly on $\mathcal{A} - \{P\}$, it follows that $\mathcal{A}' = \mathcal{A}$.

Therefore, $|N_G(R) : N_G(P)| = |\mathcal{A}| = p^m$. By (10), it follows that $m = 1$ and $|G|_p = |N_G(R)|_p = p^3$.

In $N_G(R)/R$, we see that $\overline{C_G(P)} = \overline{C_Q(P)} \rtimes \overline{C_W(P)}$, that $\overline{C_Q(P)}$ acts regularly on $\mathcal{A} - \{P\}$ and that $\overline{C_W(P)}$ acts faithfully on $\overline{C_Q(P)}$. Proposition 1 of Appendix II can then be applied to $N_G(R)/R$ acting on $\mathcal{A}$ so as to obtain

$$N_G(R)/R = (R_1/R) \rtimes C_Q(P)C_W(P),$$

where R_1 is a Sylow p-subgroup of G and $C_Q(P)$ acts regularly on $(R_1/R)^{\#}$.

As R_1 is non-abelian of order p^3, R_1 is of class $2 < p$ and so, by the Hall-Wielandt Theorem, $G/O^p(G) = N_G(R_1)/O^p(N_G(R_1))$.

Since $N_G(R)$ acts transitively on $\mathcal{A}$, $T \triangleleft R_1$; since R_1/T is abelian and $C_{R_1/R}(s) = 1$, $R_1/T = (R/T) \times (T_1/T)$, where $T_1 = [R_1/T, s]$. Let $\mathcal{A}_1$ denote the set of subgroups of R_1/T of order p which are distinct from T_1/T. As in (11) we see that $C_Q(P)$ acts regularly on $\mathcal{A}_1 - \{R/T\}$, and so, as above, we see that, if $N_G(R) \subsetneq N_G(R_1)$, then $N_G(R_1)$ acts transitively on $\mathcal{A}_1$ and
$$|N_G(R_1) : N_G(R)| = |\mathcal{A}_1| = p,$$
which contradicts (10). Thus $N_G(R_1) = N_G(R)$. Since T_1 is normalized by $C_Q(P)C_W(P)$ and by P, $T_1 C_Q(P)C_W(P)$ is normal and of index p in $N_G(R_1)$ and so hypothesis (B2) is false. $\qquad\square$

We now finish the proof in case (10.2). For a possible alternative proof in this case, see the remark at the end of §3 of Chapter IV.

Put $\Sigma = C_W(P)$ and $Z_1 = \langle st \rangle$.

(13) $C_G(Z_1)$ *is a 3-group.*

Proof. By Chapter I, §1, Lemma and Chapter I, §3, Lemma 3, $|C_G(Z_1)| = |C_G(Z_1) \cap C_G(s)| \cdot |J|$, where $J = \{x \in C_G(Z_1) \mid x^s = x^{-1}\}$. Now

$$C_G(Z_1) \cap C_G(s) = C_G(t) \cap C_G(s) = V = WP.$$

It is thus sufficient to show that $|J|$ is a power of 3. Let r be a prime divisor of $|J|$. Thus J contains an element x of order r; as x is strongly real, x is conjugate in G to an element of $\langle Q_0, K, t \rangle \cong \mathrm{PSL}(2,8)$ (Chapter I, §3, Lemmas 3 and 4). As $|\mathrm{PSL}(2,8)| = 8 \cdot 9 \cdot 7$, $r = 3$ or 7. But, if $r = 7$, then x is conjugate to an element of $K^\#$ and $C_G(x)$ is conjugate to $C_G(K)$. By Chapter I, §2, Proposition 1(a), the elements of $K^\#$ have only two fixed points so that $C_G(K) = C_D(K) = KW$. It follows that x cannot centralize a strongly real element of order 3. $\qquad\square$

(14) *In the notation of* (11), $Z(R\Sigma) = Z_1 P$. *There is a 3-subgroup R_1 of G such that $N_G(R\Sigma)/R\Sigma = (R_1/R\Sigma) \rtimes \langle s \rangle \cong S_3$. Let R_2 be a Sylow 3-subgroup of G containing R_1. Then $|R_2 : R_1| = 1$ or 3, $Z_1 = Z(R_1) = Z(R_2)$ and $R_2 = C_G(Z_1)$.*

(Here S_3 denotes the symmetric group of degree 3.)

Proof. Note first that $Z_1 P \subset Z(R\Sigma)$ (note that $Z_1 \subset T$), that $|R\Sigma/Z_1 P| = 9$ and that $R\Sigma$ is not abelian, and so $Z_1 P = Z(R\Sigma)$. Since $Z_1 = [R\Sigma, R\Sigma]$, $N_G(R\Sigma)$ acts on the set $\mathcal{A}_2$ of subgroups of order 3 in $Z_1 P$ which are distinct from Z_1. But $\mathcal{A}_2 \subset \mathcal{A}$ so that, by (11), $\langle s \rangle$ acts regularly on $\mathcal{A}_2 - \{P\}$. By (10), $R\Sigma$ is not a Sylow 3-subgroup of G and so $N_G(R\Sigma) \not\subset N_G(P)$. It follows that $N_G(R\Sigma)$ induces the group of all permutations of $\mathcal{A}_2$, a group isomorphic to S_3. The kernel of the action of $N_G(R\Sigma)$ on $\mathcal{A}_2$ is $R\Sigma$ since the kernel is contained in $N_G(P) = RC_Q(P)\Sigma$ and the elements of $C_Q(P)^\#$ act fixed-point-freely on $\mathcal{A} - \{P\}$. The existence of R_1 then follows from the structure of S_3.

By (10), $|R_2 : R_1| = 1$ or 3. We see that $Z(R_1) \subset C_{R_1}(P) = R\Sigma$ and so $Z(R_1) \subset Z(R\Sigma) = Z_1 P$. By the transitivity of R_1 on $\mathcal{A}_2$, $Z(R_1) = Z_1$. Now $Z(R_2) \subset R_2 \cap C_G(P) \subset R_1$ whence $Z(R_2) = Z(R_1) = Z_1$. We then conclude that $R_2 \subset C_G(Z_1)$ and, by (13), that $R_2 = C_G(Z_1)$. $\qquad\square$

(15) *There is a subgroup L of R_1 which is cyclic of order 9, inverted by s, normalized by V and centralized by W but not by P. It is also the case that $|R_2 : LV| = 3$, $Z(LV) = Z_1\Sigma$ and $\Omega_1(LV) = Z_1\Sigma P$.*

Proof. Let $L = C_G(st) \cap \langle Q_0, K, t \rangle$. Since $\langle Q_0, K, t \rangle \cong \mathrm{PSL}(2, 8)$, L is cyclic of order 9 and its elements are inverted by s. Now $L \subset \langle Q_0, K, t \rangle \subset C_G(W)$. As P normalizes $\langle Q_0, K, t \rangle$ and centralizes st, P normalizes L but does not centralize it due to the structure of $C_G(P)$ ($T\Sigma$ has exponent 3). Because $|LP : Z_1 P| = 3$, L normalizes $Z_1 P$ and so L normalizes

$$C_G(Z_1 P) = C_G(P) \cap C_G(st) = R\Sigma,$$

whence $L \subset R_1$. By (10), $|R_2 : LV| = 3$. We also have $Z_1\Sigma \subset Z(LV)$ and $Z(LV) \subset LW$ since LV is non-abelian, whence
$$Z(LV) \subset C_{LW}(P) = Z_1\Sigma.$$
Since $LV/Z_1\Sigma$ is abelian, $\Omega_1(LV)$ is the set of elements of order ≤ 3 in LV ([H], Kapitel III, Hilfsatz 1.3 b)). But $Z_1\Sigma P \subset \Omega_1(LV)$ and $\Omega_1(LW) = Z_1\Sigma$ whence $\Omega_1(LV) = Z_1\Sigma P$. $\qquad\square$

(16) $Z_1 P\Sigma \subset Z_2(R_1)$, Z_1 *is the only subgroup of order* 3 *in* $Z_1 P\Sigma$ *consisting of strongly real elements and* $N_G(Z_1 P\Sigma) = N_G(Z_1) = R_2\langle s\rangle$.

Proof. By (14), $Z(R_1) = Z_1 \subset Z_1 P = Z(R\Sigma) \vartriangleleft R_1$ and so $Z_1 P \subset Z_2(R_1)$. By (15), $Z_1\Sigma = Z(LV) \vartriangleleft R_1$ and so $Z_1\Sigma \subset Z_2(R_1)$. Let X be a subgroup of order 3 in $Z_1 P\Sigma$ consisting of strongly real elements such that $X \cap Z_1 = 1$. Since $Z(R_1) = Z_1 \subset Z_1 X \subset Z_2(R_1)$, R_1 permutes transitively the subgroups of order 3 in $Z_1 X$ which are distinct from Z_1. Thus, the elements of $Z_1 X$ are strongly real, and this gives a contradiction because $(Z_1 X) \cap (P\Sigma) \neq 1$.

Thus we see that $N_G(Z_1 P\Sigma) \subset N_G(Z_1) = C_G(Z_1)\langle s\rangle$ and $C_G(Z_1) = R_2$ by (14). On the other hand, $Z_1 P\Sigma = \Omega_1(LV)$ is normal in $R_2\langle s\rangle$. $\qquad\square$

(17) *Conclusion.*

Proof. Let $x \in G$ be such that $(Z_1 P\Sigma)^x \subset R_2$.

Suppose that $Z_1^x \not\subset LV$. Then $R_2 = (LV) \rtimes Z_1^x$ and $(Z_1 P\Sigma)^x = A \times Z_1^x$ where A is a subgroup of LV of type $(3, 3)$. Now $A \subset \Omega_1(LV) = Z_1 P\Sigma$. By (16), the elements of $A^\#$ are not strongly real, and so $A \cap Z_1 = 1$ and $Z_1\Sigma P = Z_1 A$. It follows that Z_1^x centralizes $Z_1\Sigma P$, whence $Z_1\Sigma \subset Z(R_2)$, a contradiction to (14).

Therefore, $Z_1^x \subset \Omega_1(LV) = Z_1\Sigma P$, whence $Z_1^x = Z_1$ and $x \in N_G(Z_1) = R_2\langle s\rangle$ by (16), and so x normalizes $Z_1 P\Sigma$.

The subgroup $Z_1 P\Sigma$ of G is thus weakly closed in R_2 and, as $Z_1 P\Sigma$ is abelian, we obtain $G/O^3(G) \cong R_2\langle s\rangle/O^3(R_2\langle s\rangle)$ by the Hall-Wielandt Theorem.

If $\overline{R_1} = R_1/Z_1$, $\overline{R_1}$ is generated by the subgroups $\overline{R\Sigma} = \overline{T} \times \overline{P} \times \overline{\Sigma}$ and $\overline{L\Sigma P} = \overline{L} \times \overline{\Sigma} \times \overline{P}$ since $L \not\subset R\Sigma$. Thus $[\overline{R_1}, \overline{R_1}]$ is the normal subgroup of $\overline{R_1}$ generated by a commutator $[\overline{x}, \overline{y}]$ with $\overline{x} \in \overline{T}^\#$ and $\overline{y} \in \overline{L}^\#$. But $\overline{R_1}$ has class ≤ 2 since $Z_1 P\Sigma \subset Z_2(R_1)$ and $|R_1 : Z_1 P\Sigma| = 9$. Therefore $[\overline{x}, \overline{y}] \in Z(\overline{R_1})$ and $[\overline{R_1}, \overline{R_1}]$ has order 1 or 3. As s centralizes $\overline{\Sigma P}$, $\overline{R_1}/[\overline{R_1}, \overline{R_1}]$ then has a subgroup of order 3 centralized by s, whence $(\overline{R_1}/[\overline{R_1}, \overline{R_1}]) \rtimes \langle s\rangle$ has a quotient of order 3. If $|W| = 3$, then $R_2 = R_1$, a contradiction to hypothesis (B2).

Thus $|W| = 9$. By (14), s inverts the elements of $R_1/R\Sigma$; it follows that $C_{R_1}(s) \subset R\Sigma$ and $C_{R_1}(s) = P\Sigma$, whence $W \not\subset R_1$. Thus $R_2 = R_1 W$, and so s centralizes R_2/R_1 and $R_1\langle s\rangle \vartriangleleft R_2\langle s\rangle$. This is again a contradiction to hypothesis (B2). $\qquad\square$

Chapter III. The Structure of H

§ 1. The Structure of Q

Suppose that $V = 1$. We know that G is doubly transitive on Ω and the elements of $D^{\#}$ have only two fixed points (Chapter I, § 1, Proposition 6(c)). Furthermore, $|\Omega| = |Q| + 1$ is odd and $O_{2'}(G) = 1$ by (A2), (A3) and Chapter I, § 1, Proposition 1(e), whence G has no normal subgroup which is regular on Ω. Thus G is a Zassenhaus group. By [HB], Chapter XI, Theorem 11.16, G is isomorphic to $\mathrm{PSL}(2, q)$ or to $\mathrm{Sz}(q)$, and the conclusion of Theorem A is valid. Taking Theorem B into account, **we will then assume from this point on that**

(C1) *The subgroup V is non-trivial; $C_G(P)$ has 2-rank ≥ 2 for every subgroup P of V which is of prime order.*

Theorem C. Q *is a 2-group.*

Proof. Suppose that $Q_1 \neq 1$. Let P be a subgroup of D of prime order. If $|\Omega_P| = 2$, then $C_H(P) \subset D$ whence $C_Q(P) = 1$. If $|\Omega_P| \geq 3$, then P is conjugate in D to a subgroup of V (Chapter I, § 1, Proposition 6(c)), whence $C_{Q_1}(P) = 1$ by (C1) and Chapter I, § 3, Proposition 1(c). Thus D acts fixed-point-freely on Q_1. Moreover, $Q \cap Q^x \subset Q \cap (H \cap H^x) = 1$ for $x \in G - H$ since $H \cap H^x$ is conjugate to D in H. The hypotheses of the Feit-Sibley Theorem, stated in Appendix IV, are therefore satisfied. By this theorem, $S = \{\chi \in \mathrm{Irr}(H) \mid Q_1 \not\subset \mathrm{Ker}\,\chi\}$ is coherent for Ind_H^G.

Let λ be a linear character of H, $\lambda \neq 1_H$, such that $QK \subset \mathrm{Ker}\,\lambda$. There is such a character since $H/QK \cong V$ is solvable and non-trivial.

If $(|K|, |V|) \neq 1$, then D has a non-cyclic subgroup of order p^2, p prime, and so D cannot act fixed-point-freely on Q_1 ([H], Kapitel V, Satz 8.15). Thus QK is a Hall subgroup of H.

Let $x \in H$ and $g \in G$ be such that $x^g \in H$. We show that $\lambda(x^g) = \lambda(x)$. If π is the set of prime divisors of $|QK|$ and y is the π'-component of x, then $\lambda(x) = \lambda(y)$ and $\lambda(x^g) = \lambda(y^g)$ since the π-component of x is in QK. Thus we may assume that x is a π'-element. Therefore x and x^g are conjugate in H to elements of V by a theorem of Hall, and we may assume that $x \in V$ and $x^g \in V$. By Chapter I, § 3, Lemma 2, x and x^g are conjugate in V and so $\lambda(x) = \lambda(x^g)$.

It follows from the definition of Ind_H^G that, for $x \in H$,

$$(\mathrm{Ind}_H^G \lambda)(x) = \lambda(x)(\mathrm{Ind}_H^G 1_H)(x),$$

whence

$$
\begin{aligned}
(\mathrm{Ind}_H^G \lambda, \mathrm{Ind}_H^G \lambda) \;&=\; (\mathrm{Res}_H^G \mathrm{Ind}_H^G \lambda, \lambda) \;=\; (\lambda \mathrm{Res}_H^G \mathrm{Ind}_H^G 1_H, \lambda) \\
&=\; (\mathrm{Res}_H^G \mathrm{Ind}_H^G 1_H, 1_H) \;=\; (\mathrm{Ind}_H^G 1_H, \mathrm{Ind}_H^G 1_H) \;=\; 2
\end{aligned}
$$

since $|\lambda(x)| = 1$ for $x \in H$ and since G is doubly transitive on Ω.

Set $\mathrm{Ind}_H^G \lambda = f_1 + f_2$ with $f_i \in \mathrm{Irr}(G)\,(i = 1, 2)$. Then $(\mathrm{Ind}_H^G \lambda, 1_G) = (\lambda, 1_H) = 0$, and so $f_i \neq 1_G$.

Let $\mathcal{S} = \{\chi_1, \ldots, \chi_n\}$, with $\chi_i(1) = a_i|D|$ and $a_1 = 1$, and let $e_i \in \pm\mathrm{Irr}(G)$ be such that $\mathrm{Ind}_H^G(\chi_i - a_i\chi_1) = e_i - a_i e_1$ for $i \geq 2$; the coherence of $\mathcal{S}$ makes this possible.

Suppose that $f_1 = \pm e_i$ for some i. By Lemma 2(c) of Appendix IV, $\overline{\chi_i} \neq \chi_i$ and $\overline{\chi_i} \in \mathcal{S}$. Therefore there is an element $e_i' \in \{e_j \mid j \neq i\}$ such that $\mathrm{Ind}_H^G(\chi_i - \overline{\chi_i}) = e_i - e_i'$. By [Is], Lemma 7.7, $\mathrm{Res}_H^G(e_i - e_i') = \chi_i - \overline{\chi_i}$ since Q is a Hall subgroup of H and $\chi_i - \overline{\chi_i}$ vanishes on $H - Q$. Then

$$(\mathrm{Ind}_H^G \lambda, e_i - e_i') = (\lambda, \chi_i - \overline{\chi_i}) = 0,$$

whence $\mathrm{Ind}_H^G \lambda = \pm(e_i + e_i')$ and $|Q| + 1 = (\mathrm{Ind}_H^G \lambda)(1) = \pm 2e_i(1)$, which is impossible since $|Q|$ is even.

It follows that, for $j = 1, 2$ and $i \geq 2$, $(f_j, e_i - a_i e_1) = 0$, whence

$$(\mathrm{Res}_H^G f_j, \chi_i - a_i\chi_1) = 0.$$

Thus there are integers $b_j \in \mathbf{N}$ and characters ψ_j of H such that $Q_1 \subset \mathrm{Ker}\,\psi_j$ and $\mathrm{Res}_H^G f_j = b_j(\sum a_i\chi_i) + \psi_j$. Then

$$\begin{aligned}
|Q| + 1 = f_1(1) + f_2(1) &\geq (b_1 + b_2)\sum a_i\chi_i(1) \\
&= (b_1 + b_2)(|H| - |H/Q_1|)/|D| \\
&= (b_1 + b_2)|\mathcal{S}|(|Q_1| - 1).
\end{aligned}$$

Thus $(b_1 + b_2)(|Q_1| - 1) \leq |Q_1|$ and so $b_1 + b_2 \leq |Q_1|/(|Q_1| - 1) < 2$. It follows that there is an index $j \in \{1, 2\}$ such that $b_j = 0$ and so $Q_1 \subset \mathrm{Ker}\,f_j$.

Therefore $N = \mathrm{Ker}\,f_j$ is a normal subgroup of G such that $1 \neq N \neq G$. By Chapter I, §3, Proposition 2, G satisfies the conclusion of Theorem A and so $Q_1 = 1$ (Chapter I, §3, Lemma 1).   □

Proposition. *One of the following three cases holds.*

   (a) *$S = Q_0$ and st has order 3.*

   (b) *S is a Suzuki 2-group of type A, st has order 5 and $W = 1$.*

   (c) *S is a Suzuki 2-group of type B, st has order 3 and $W \neq 1$.*

Proof. Let P be a subgroup of V of prime order p; if $W \neq 1$, assume that $P \subset W$. Let $F = O^{2'}(C_G(P))$ and $\ell = |C_{Q_0}(P)|$. By (C1) and Chapter I, §3, Proposition 1, one of the following three cases holds.

   st has order 3, $C_S(P)$ is elementary abelian, $F/Z(F) \cong \mathrm{PSL}(2, \ell)$.

   st has order 5, $C_S(P)$ is a Suzuki 2-group of type A, $F/Z(F) \cong \mathrm{Sz}(\ell)$.

   st has order 3, $C_S(P)$ is a Suzuki 2-group of order ℓ^3, $F/Z(F) \cong \mathrm{PSU}(3, \ell)$.

We know that either S is abelian or S is a Suzuki 2-group (Chapter I, §2). We make use of the definitions and of the theorem of Appendix III which provides an exposition of Suzuki 2-groups.

(1) <u>Suppose that S is abelian.</u>

Then $C_S(P)$ is abelian and so st has order 3 and $C_S(P) \subset Q_0$. Suppose that $S \neq Q_0$. There is then an element $x \in S$ such that $x^2 = s$ (since K is transitive on $Q_0^{\#}$); since S is abelian, $\{y \in S \mid y^2 = s\} = xQ_0$. But P centralizes s (Chapter I, §1, Proposition 5) and so normalizes xQ_0 which is of cardinality prime to p, whence $C_S(P) \not\subset Q_0$, which is a contradiction. Thus $S = Q_0$.

(2) <u>Suppose that S is non-abelian of order q^2.</u>

In this case S is a Suzuki 2-group of type A. Let $x \in S$ be such that $x^2 = s$. Since $|\{y \in S \mid y^2 = s\}| = (q^2 - q)/(q - 1) = q$, we again find that $\{y \in S \mid y^2 = s\} = xQ_0$ and that P normalizes xQ_0 whence $C_S(P)$ has exponent 4. If $W \neq 1$, then $C_S(P)$ is a K-subgroup of S which has exponent 4 and so $C_S(P) = S$, which is a contradiction since D acts faithfully on S. Thus, $W = 1$. By Chapter I, §2, Proposition 3, V then acts as a group of field automorphisms on Q_0 and, by the theorem of Galois, $C_V(C_{Q_0}(P)) = P$.

But, if $G_0 = \mathrm{PSU}(3, \ell)$, S_0 is a Sylow 2-subgroup of G_0 and $N_{G_0}(S_0) = S_0 \rtimes D_0$, then, as can be checked, $C_{D_0}(\Omega_1(S_0)) \neq 1$.

It follows that $F/Z(F)$ is not isomorphic to $\mathrm{PSU}(3, \ell)$ and so, since $C_S(P)$ has exponent 4, $F/Z(F) \cong \mathrm{Sz}(\ell)$ and st has order 5.

(3) <u>Suppose that S is non-abelian of order q^3.</u>

If S is a Suzuki 2-group of type C or type D, then S/Q_0 is an $\mathbf{F}_2[K]$-module satisfying $S/Q_0 = X \oplus Y$, where X and Y are non-isomorphic $\mathbf{F}_2[K]$-modules of order q. It follows that X and Y are the only $\mathbf{F}_2[K]$-submodules of order q in S/Q_0. As P acts on $(S/Q_0) \rtimes K$, P therefore normalizes X and Y.

Suppose that st has order 5. It follows that $C_S(P)$ is a Suzuki 2-group of type A. If S is of type B, then every element of order 4 in S generates a K-subgroup of order q^2, and the number of K-subgroups of S of order q^2 is $q + 1$. Since P centralizes an element of order 4 in S, it follows that P normalizes at least two K-subgroups X and Y of order q^2 in S. By the preceding paragraph, this remains true if S is of type C or type D. As in case (2), P then centralizes an element $x \in X$ and an element $y \in Y$ such that $x^2 = y^2 = s$. But, since $C_S(P)$ is of type A, it follows that $y \in x\Omega_1 C_S(P)$ and $y \in X$, which is a contradiction. Thus st has order 3.

Suppose that $W = 1$. Then, as in case (2), $C_V(C_{Q_0}(P)) = P$, $F/Z(F)$ is not isomorphic to $\mathrm{PSU}(3, \ell)$ and the case in which $C_S(P)$ is elementary abelian holds since st has order 3. But $[K, P] \rtimes P$ is a Frobenius group acting on S/Q_0 and $[K, P]$ acts without fixed points on S/Q_0, whence $C_{S/Q_0}(P) \neq 1$ and we obtain a contradiction. Thus, $W \neq 1$.

By Chapter I, §3, Lemma 5, S is then of type B. $\qquad\qquad\square$

§ 2. The Case in which st has Order 5

Proposition. *If case* (b) *of the proposition of* § 1 *holds, then* $(SK) \cup (SKtS)$ *is a subgroup of* G.

Proof. It suffices to show that $tSt \subset SKtS$. Let $f, g : S^\# \to S^\#$ and $h : S^\# \to D$ be the mappings such that, for $x \in S^\#$, $txt = g(x)h(x)tf(x)$. The existence of these mappings follows from the fact that $txt \notin H \cup (Ht) \cup (tH)$ and from Chapter I, § 1, Proposition 4(a). It is thus sufficient to show that $h(x) \in K$ for all $x \in S^\#$.

For $a \in K$, putting $tx^a t = atxta^{-1}$ into canonical form, we see that

(1) $f(x^a) = f(x)^{a^{-1}}$, $g(x^a) = g(x)^{a^{-1}}$ and $h(x^a) = ah(x)a$ $(x \in S^\#, a \in K)$.

It is thus sufficient to show that $h(x) \in K$ for all elements x in a system of representatives for the K-orbits of $S^\#$. Let

(2) $tst = r^{-1}tr$

be the structure equation of G (Chapter I, § 1, Proposition 4). Then $(st)^2 = (st)^r$ and, as st has order 5, $(st)^{r^2} \neq st$, whence $r^2 \neq 1$ and r has order 4.

By (2),

(3) $trt = rts$ and $tr^{-1}t = str^{-1}$.

In particular, $h(s) = h(r) = h(r^{-1}) = 1$. Let $k \in K - \{1\}$, and let ℓ be that element of K for which $sksk^{-1} = s^\ell$. Then

$$
\begin{aligned}
trr^{-k}t &= trt \cdot tr^{-k}t = rts \cdot k \cdot str^{-1} \cdot k^{-1} \\
&= rt \cdot sksk^{-1} \cdot tk^{-2}r^{-k^{-1}} = r\ell \cdot r^{-1}tr \cdot \ell^{-1}k^{-2}r^{-k^{-1}} \\
&= rr^{-\ell^{-1}}\ell^2 k^2 tr^{\ell^{-1}k^{-2}} r^{-k^{-1}},
\end{aligned}
$$

which is to say that

(4) $f(rr^{-k}) = r^{\ell^{-1}k^{-2}} r^{-k^{-1}}$, $g(rr^{-k}) = rr^{-\ell^{-1}}$ and $h(rr^{-k}) = \ell^2 k^2$.

In particular, $h(rr^{-k}) \in K$. It is thus sufficient to show that s, r, r^{-1} and the elements rr^{-k} for $k \in K^\#$ form a system of representatives for the K-orbits of $S^\#$, or, since $|S^\#|/|K| = q+1 = |K^\#|+3$, that these elements are pairwise non-conjugate under the action of K.

First of all, as r has order 4 and $|K|$ is odd, s, r and r^{-1} are in pairwise distinct K-orbits. If $k \in K^\#$ and $rr^{-k} = z \in Q_0$, then $(r^2)^k = (r^k)^2 = (rz)^2 = r^2$, which is impossible. Therefore, rr^{-k} has order 4. Thus, rr^{-k} is not K-conjugate to s for all $k \in K^\#$.

It follows that, for $k \in K^\#$, $f(rr^{-k}) = (rr^{-k\ell})^{\ell^{-1}k^{-2}}$ and $g(rr^{-k}) = rr^{-\ell^{-1}}$ have order 4, whence, by (1) and (3), rr^{-k} is not K-conjugate to either r or r^{-1}. It remains to show that the elements rr^{-k} are in pairwise distinct K-orbits for $k \in K^\#$.

Since K acts regularly on $(S/Q_0)^{\#}$, we can identify S/Q_0 with $\mathbf{F}_q$ and K with $\mathbf{F}_q^*$ in such a way that the action of K on S/Q_0 is identified with the action of $\mathbf{F}_q^*$ on $\mathbf{F}_q$ by multiplication (Appendix I, Proposition 2). Let α be the canonical surjection from S onto $\mathbf{F}_q$.

Let $a \in K$ and $k_1, k_2 \in K^{\#}$ be such that $rr^{-k_2} = (rr^{-k_1})^a$. By use of (1) and (4), applying α to the identities $rr^{-k_2} = (rr^{-k_1})^a$, $h(rr^{-k_2}) = h((rr^{-k_1})^a)$ and $f(rr^{-k_2}) = f((rr^{-k_1})^a)$, we obtain

(5) $1 + k_2 = a(1 + k_1)$,

(6) $\ell_2 k_2 = a\ell_1 k_1$,

(7) $\ell_2^{-1} k_2^{-2} + k_2^{-1} = a^{-1}(\ell_1^{-1} k_1^{-2} + k_1^{-1})$,

where $sk_i sk_i^{-1} = s^{\ell_i}$ $(i = 1, 2)$. Thus, on dividing (5) by (6) term by term and on multiplying (6) by (7), we obtain $x_1 = x_2$ and $y_1 = y_2$, where $x_i = \ell_i^{-1}(k_i^{-1} + 1)$ and $y_i = k_i^{-1} + \ell_i$. But $(x_i + 1)k_i^{-1} = x_i y_i + 1$, whence $k_1 = k_2$ provided that $x_i \neq 1$.

We show that $\ell^{-1}(k^{-1} + 1) \neq 1$ for $k \in K^{\#}$ and ℓ such that $sksk^{-1} = s^{\ell}$. If not, $\ell = k^{-1} + 1$ and

$$\alpha(f(rr^{-k})g(rr^{-k}))$$
$$= \ell^{-1}k^{-2} + k^{-1} + 1 + \ell^{-1} = (k^{-1} + 1)(1 + \ell^{-1}(k^{-1} + 1)) = 0.$$

Therefore, putting $f = f(rr^{-k})$, $g = g(rr^{-k})$ and $h = h(rr^{-k})$, we obtain $fg \in Q_0$ and $(trr^{-k}t)^2 = ghtfghtf = gt(fg)^h tf$.

As rr^{-k} has order 4, $fg \neq 1$ and $t(fg)^h tfg$ is an involution. This is a contradiction since $t(fg)^h t \in I \cap (G - H)$ and $fg \in I \cap H$. $\qquad\square$

§3. The Action of KW on S

Assume that case (a) or (b) of the proposition in §1 holds. Then $G_0 = (SK) \cup (SKtS)$ is a subgroup of G (§2 and Chapter I, §3, Lemma 4). Also, $G = H \cup (HtS) = \langle G_0, V \rangle$ and V normalizes S, K and t whence $G_0 \triangleleft G$ and $|G/G_0| = |V|$. The conclusion of Theorem A now follows from Chapter I, §3, Proposition 2. **Therefore, we assume henceforth that**

(C2) *S is a Suzuki 2-group of type B, st has order 3 and $W \neq 1$.*

Let $F = \mathbf{F}_q$ and $E = \mathbf{F}_{q^2}$. Let $S \cong B(n, \theta, \varepsilon)$ for some odd order automorphism θ of F. As described in Appendix III, Definition 3, S is a central extension

$$F \to S \to F \times F$$

with associated quadratic mapping $\chi : F \times F \to F$ given by

$$\chi(a, b) = a^{1+\theta} + \varepsilon ab^{\theta} + b^{1+\theta}$$

and with $\varepsilon \in F$ such that $\chi(a, b) = 0$ implies that $a = b = 0$. Furthermore, K can be identified with F^* in such a way that the actions of K on S/Q_0 and on Q_0, identified with $F \times F$ and with F respectively, are given by

$$(a, b)^x = (xa, xb) \quad \text{and} \quad c^x = x^{1+\theta}c \quad (x \in F^*, \ (a, b) \in F \times F, \ c \in F).$$

For $x \in E$, we set $\overline{x} = x^q$.

Proposition. *There is an isomorphism from $S \rtimes KW$ to $S_1 \rtimes K_1W_1$, which sends S onto S_1, K onto K_1, and s to $(0, 1)$, where $S_1 \rtimes K_1W_1$ satisfies the following conditions.*

S_1 is the set of all pairs $(x, y), x, y \in E$, with $y \in F$ if $\theta \neq 1$ and $y + y^q = x^{1+q}$ if $\theta = 1$. The operation in S_1 is

$$(x, z)(y, u) = (x + y, z + u + \varphi(x, y)).$$

If $\theta = 1$, $\varphi(x, y) = xy^q$. Otherwise, φ is a biadditive mapping $E \times E \to F$ such that $\varphi(ax, by) = ab^\theta \varphi(x, y)$ for $a, b \in F$ and such that $\varphi(x, x) \neq 0$ if $x \neq 0$.

K_1W_1 is a subgroup of E^* with $K_1 = F^*$ and W_1 a non-trivial subgroup of $\{x \in E^* \mid x^{1+q} = 1\}$. There is an automorphism σ of E whose restriction to F is θ and which satisfies $x^\sigma = x^{-1}$ for $x \in W_1$. The action of K_1W_1 on S_1 is given by $(x, y)^a = (ax, a^{1+\sigma}y)$ for $a \in K_1W_1$.

Proof.

(1) <u>Identification of $(S/Q_0) \rtimes KW$ with $E \rtimes K_1W_1$.</u>

By Chapter I, §3, Lemma 5, W is cyclic of order dividing $q + 1$ and acts fixed-point-freely on the set of K-subgroups of S/Q_0. Let w be a generator of W. Then w is an F-linear automorphism of $S/Q_0 \cong F \times F$.

If $\theta = 1$, an identification of (S/Q_0) with E, compatible with its F-space structure, is made in Appendix III, Proposition 1, in such a way that the quadratic mapping χ associated with the extension $F \to S \to E$ is given by $\chi(x) = x\overline{x}$. By Proposition 2 of Appendix III, there is an element $\omega \in E^*$ and an automorphism τ of E such that, for $x \in S/Q_0 \cong E$, $x^w = \omega x^\tau$ and, for $y \in Q_0 \cong F$, $y^w = \omega\overline{\omega}y^\tau$. Since w acts trivially on Q_0, $\omega\overline{\omega} = 1$ and $x^\tau = x$ or $x^\tau = \overline{x}$. But, in the second case, w would have even order and so $x^w = \omega x$ for $x \in E$.

If $\theta \neq 1$, the characteristic polynomial $P(T)$ of w as F-automorphism of S/Q_0 is irreducible since w does not stabilize any subspace of dimension 1 in S/Q_0; thus, S/Q_0 can be identified in an F-linear manner with $F[T]/(P) \cong E$, with the operation of w on S/Q_0 being identified with multiplication by an element ω of E such that $P(\omega) = 0$.

(2) <u>Existence of σ.</u>

If $\theta = 1$ and $x^\sigma = x^q$, then $\sigma_{|F} = \theta$ and $x^\sigma = x^{-1}$ for $x \in W_1$. Assume that $\theta \neq 1$. Let $\chi : E \to F$ be the quadratic mapping associated with the central extension $F \to S \to E$. Let $\lambda_{\mu\nu} \in E$ be such that $\chi(x) = \sum \lambda_{\mu\nu} x^\mu x^\nu$, where the sum is taken over all subsets $\{\mu, \nu\}$ of $\mathrm{Aut}(E)$ of cardinality 1 or 2 (Appendix III, Lemma 2(c)). Noting that $\chi(x) = \overline{\chi(x)}$ and that, for $a \in F$ and $x \in E$, $\chi(ax) = a^{1+\theta}\chi(x)$ and applying the lemma just cited, we see that $\lambda_{\overline{\mu},\overline{\nu}} = \overline{\lambda_{\mu\nu}}$, where $\overline{\mu}(x) = \overline{\mu(x)}$, and that, if $\lambda_{\mu\nu} \neq 0$, then $a^\mu a^\nu = aa^\theta$ for $a \in F$, whence $\{\mu_{|F}, \nu_{|F}\} = \{1_F, \theta\}$. It follows that, if σ is an automorphism of E which extends θ, then

$$\chi(x) = \lambda_1 x^{1+\sigma} + \overline{\lambda_1}\overline{x}^{1+\sigma} + \lambda_2 x^{q+\sigma} + \overline{\lambda_2}\overline{x}^{q+\sigma}$$

with $\lambda_1, \lambda_2 \in E$ not both zero. Since w acts trivially on Q_0, $\chi(wx) = \chi(x)$ and so, by the same lemma, $\omega^{1+\sigma} = 1$ if $\lambda_1 \neq 0$ and $\omega^{q+\sigma} = 1$ if $\lambda_2 \neq 0$. As $\omega^q \neq \omega$, we obtain $\lambda_2 = 0$ and $\omega^{1+\sigma} = 1$, possibly on replacing σ by $\overline{\sigma}$.

(3) <u>Identification of S with S_1.</u>

$$\varphi(x,y) = \begin{cases} xy^q & \text{if } \theta = 1, \\ \lambda_1 xy^\sigma + \overline{\lambda_1}\overline{xy^\sigma} & \text{if not.} \end{cases}$$

It can then be checked that the operation given in the statement makes S_1 a group and that $F \xrightarrow{\iota} S_1 \xrightarrow{\pi} E$ (with $\iota(y) = (0,y)$ and $\pi(x,y) = x$) is a central extension whose associated quadratic mapping is χ. By Lemma 1(c) of Appendix III, this extension is equivalent to the extension $F \to S \to E$.

(4) <u>Action of $K_1 W_1$ on S_1.</u>

Let A be the image of KW in $\mathrm{Aut}(S_1)$ where KW acts by conjugation on $S \cong S_1$. It can be checked that the formula given in the statement defines an action of $K_1 W_1$ on S_1. Let B be the image of $K_1 W_1$ in $\mathrm{Aut}(S_1)$ for this action. Let U be the group of automorphisms of S_1 which induce the identity on $Z(S_1)$ and on $S_1/Z(S_1)$. Then $U \triangleleft \mathrm{Aut}(S_1)$ and, by (1), $B \subset UA$. But U is a 2-group (Appendix III, Lemma 1(d)) and so, by a theorem of Zassenhaus ([H], Kapitel I, Hauptsatz 18.3), there is an element $u \in U$ such that $A^u = B$, and u induces an isomorphism from $S_1 \rtimes A$ to $S_1 \rtimes B$.

(5) <u>Conclusion.</u>

Since K is transitive on $Q_0^{\#}$, we can assume that s is identified with $(0,1)$ on composing the isomorphism $S \rtimes KW \to S_1 \rtimes K_1 W_1$ with an inner automorphism of $S_1 \rtimes K_1 W_1$. $\qquad\square$

Chapter IV. Characterization of $\mathrm{PSU}(3, q)$

§ 1. The Mappings f, g and h

In this chapter we conclude the proof of Theorem A. We show in particular that, if $V = W$, then G is isomorphic either to $\mathrm{PSU}(3, q)$ or to $\mathrm{PGU}(3, q)$. The proof is accomplished by an explicit calculation with the operation in G; it follows a method due to Suzuki.

Suppose that L is a finite group acting doubly transitively on a set X, and that M is the stabilizer in L of a point in X, t is an involution in $L - M$ and $D = M \cap M^t$. We also assume that there is a subgroup Q of M such that $M = Q \rtimes D$ (these hypotheses mean that L has a split BN-pair of rank 1).

Then there are uniquely determined mappings

$$f, g : Q^\# \to Q^\# \quad \text{and} \quad h : Q^\# \to D$$

such that, for $x \in Q^\#$,

$$txt = g(x)h(x)tf(x).$$

Indeed, as in Chapter I, §1, Proposition 4, we see that every element of $L - M$ can be expressed uniquely as atb with $a \in M$ and $b \in Q$, and that $txt \notin M \cup (Mt) \cup (tM)$ since $Q^t \cap M = 1$.

These mappings satisfy the following identities, in which the x denotes an arbitrary element of $Q^\#$.

(H1) $f(x^{-1}) = g(x)^{-1}$.

(H2) $f(f(x)) = x$.

(H3) $f(x^a) = f(x)^{a^t}$ *for* $a \in D$.

(H4) $h(x^a) = a^{-t}h(x)a$ *for* $a \in D$, $h(x^{-1}) = h(x)^{-t}$ *and* $h(f(x)) = h(x)^{-1}$.

(H5) *Let* j *be the mapping* $x \mapsto x^{-1}$ *from* $Q^\#$ *to* $Q^\#$. *Then*

$$(f \circ j)^3(x) = x^{h(x)^{-1}}.$$

(H6) *If* $x, y \in Q^\#$ *and* $xy \neq 1$, *then*

$$\begin{aligned}
f(x)g(y) &\neq 1, \\
f(xy) &= f(f(x)g(y))^{h(y)^t} f(y)
\end{aligned}$$

and

$$h(xy) = h(x)h(f(x)g(y))h(y).$$

Proof. Identities (H1) to (H4) are proved by straightforward calculations. Let $x \in Q^\#$. Then $txt = g(x)h(x)tf(x)$, whence $tf(x)t = h(x)^{-1}g(x)^{-1}tx$ and so $g(f(x)) = g(x)^{-h(x)}$. By (H1), $(j \circ f)^2(x) = ((f \circ j)(x))^{h(x)}$ and, by (H2) and (H3),

$$x = (f \circ j)^2((f \circ j)(x))^{h(x)} = ((f \circ j)^3(x))^{h(x)},$$

which proves (H5). Let $x, y \in Q^{\#}$ be such that $xy \neq 1$ and let $z = f(x)g(y)$. Then

$$txyt = g(x)h(x)tzh(y)tf(y) = g(x)h(x)tzth(y)^t f(y),$$

whence $z \neq 1$ and

$$
\begin{aligned}
txyt &= g(x)h(x)g(z)h(z)tf(z)h(y)^t f(y) \\
&= g(x)h(x)g(z)h(z)h(y)tf(z)^{h(y)^t} f(y),
\end{aligned}
$$

which proves (H6). $\qquad\qquad\square$

We note that, by (H2), (H3) and (H5), $\langle f, j \rangle$ acts on the set of orbits of $Q^{\#}$ under D, and that the permutation group induced by $\langle f, j \rangle$ on this set is isomorphic to a quotient of the dihedral group of order 6.

Lemma. *If L acts faithfully on X, then $\langle Q^x \mid x \in L \rangle$ is determined up to isomorphism by Q and f, and also L is determined up to isomorphism by $M = Q \rtimes D$ and f.*

Proof. Let a be a point in X whose stabilizer in L is M. We can identify X with $Q \cup \{a\}$ by identifying $x \in Q$ with a^{tx}. Then the action of t on X is determined by f, and the action of Q (M, respectively) is determined by the specification of Q ($Q \rtimes D$, respectively). But then we have $\langle Q^x \mid x \in L \rangle = \langle Q, Q^{tx} \mid x \in Q \rangle \subset \langle Q, t \rangle$ and $L = \langle M, t \rangle$. $\qquad\square$

In the following calculations, **we resume the hypotheses (C1) and (C2) of Chapter III,** and we work towards determining f for the case $L = G$ and $M = H$. We identify $Q \rtimes KW$ with the group $S_1 \rtimes K_1 W_1$ defined in §3 of Chapter III.

§2. Preliminary Calculation

By (C2), $tst = sts$ whence, by (H3) and (H4),

(1) *for $a \in K$, $f(s^a) = g(s^a) = s^{a^{-1}}$ and $h(s^a) = a^2$.*

By (H6) applied to $x = \omega \in Q - Q_0$ and $y = s^a$, we obtain

(2) $f(\omega s^a) = f(f(\omega)s^{a^{-1}})^{a^{-2}} s^{a^{-1}}$ *for $\omega \in Q - Q_0$ and $a \in K$.*

Applying (H6) to $x = s^a$ and $y = \omega$, we also obtain

(3) $f(\omega s^a) = f(g(\omega)s^{a^{-1}})^{h(\omega)^t} f(\omega).$

(4) *If $f(\omega x) = f(\omega)y$ for some $\omega \in Q - Q_0$ and $x, y \in Q_0$, then $x = 1$.*

Proof. If $x \neq 1$, there is an element $k \in K$ such that $x = s^k$. By (3), we then have $f(\omega)y = f(g(\omega)s^{k^{-1}})^{h(\omega)^t} f(\omega)$ and so $f(g(\omega)s^{k^{-1}}) \in Q_0$, whence $g(\omega) \in Q_0$ and then $\omega \in Q_0$, which is a contradiction. $\qquad\square$

(5) *If $f(\omega) = (\omega y)^a$ for some $\omega \in Q - Q_0$, $y \in Q_0$ and $a \in D$, then $y \neq 1$ and $a \notin K$.*

Proof. Since $|D|$ is odd, j has no fixed points on the set of orbits of $Q - Q_0$ under D, and, since f is conjugate to j in the permutation group induced by $\langle f, j \rangle$ on this set, neither does f, and so $y \neq 1$. By (H2) and (H3),

$$f(\omega y) = \omega^{a^{-t}} = (f(\omega)^{a^{-1}} y)^{a^{-t}} = (f(\omega)y^a)^{a^{-1}a^{-t}}$$

and so, by (4), $a^{-1}a^{-t} \neq 1$, whence $a \notin K$. $\qquad\square$

(6) *If $f(\omega x) = (f(\omega)y)^a$ for some $\omega \in Q - Q_0$, $x, y \in Q_0$, $x \neq 1$, and $a \in D$, then $a \notin K$.*

Proof. Let $k \in K$ be such that $x = s^k$. By (2), $(f(\omega)y)^a = f(f(\omega)s^{k^{-1}})^{k^{-2}}s^{k^{-1}}$, whence $f(f(\omega)s^{k^{-1}}) = ((f(\omega)y)^a s^{k^{-1}})^{k^2} = (f(\omega)s^{k^{-1}} \cdot s^{k^{-1}} y s^{k^{-1}a^{-1}})^{ak^2}$. By (5), it follows that $ak^2 \notin K$, whence $a \notin K$. $\qquad\square$

(7) *Let $\omega, \omega' \in Q - Q_0$ and let $x_i, y_i \in Q_0$ and $a_i \in D$ $(i = 1, 2)$ be such that $x_1 \neq x_2$ and $f(\omega x_i) = (\omega' y_i)^{a_i}$ $(i = 1, 2)$. Then $a_2 \notin a_1 K$.*

Proof. The hypothesis implies that

$$f(\omega x_2) = (f(\omega x_1)^{a_1^{-1}} y_1 y_2)^{a_2} = (f(\omega x_1)(y_1 y_2)^{a_1})^{a_1^{-1}a_2}.$$

By (6), $a_1^{-1}a_2 \notin K$. $\qquad\square$

Set $m = |W|$ and $n = (q + 1)/m = |E^*/KW|$. For $\omega \in Q$, let $\overline{\omega}$ denote the image of ω in Q/Q_0. If $\omega = (\alpha, \beta)$, then $\overline{\omega}$ is identified with $\alpha \in E$. Let $\omega_1, \dots, \omega_n \in Q - Q_0$ be such that the elements $\overline{\omega_i}$ comprise a system of representatives of the orbits of $(Q/Q_0)^\#$ under KW.

(8) *The number of elements $x \in Q_0$ such that $\overline{f(\omega_1 x)}$ is in the orbit of $\overline{\omega_i}$ under KW is m if $i > 1$ and $m - 1$ if $i = 1$.*

Proof. Let m_i be the number of $x \in Q_0$ such that $\overline{f(\omega_1 x)}$ is in the orbit of $\overline{\omega_i}$ under KW. Then $m_i \leq m$ for $i > 1$ by (7) with $\omega = \omega_1$ and $\omega' = \omega_i$. Again, by (7) and (5), $m_1 \leq m - 1$. Therefore $q = \sum m_i \leq nm - 1 = q$, whence all the inequalities are in fact equalities. $\qquad\square$

Let ζ be a generator of W. By (C2), $\zeta \neq 1$.

(9) *For all i $(1 \leq i \leq n)$, there are elements $\omega_i' \in Q - Q_0$ and $y_i \in Q_0^\#$ such that $\overline{\omega_i'}$ is in the orbit of $\overline{\omega_i}$ under KW and $f(\omega_i') = (\omega_i' y_i)^\zeta$.*

Proof. Let ω be one of the elements ω_i. By (5), (7) and (8), there are elements $x, z \in Q_0$ and $k \in K$ such that $f(\omega x) = (\omega z)^{k\zeta}$. If $a \in K$, then

$$f((\omega x)^a) = f(\omega x)^{a^{-1}} = (\omega z)^{a^{-1}k\zeta} = ((\omega x)^a(xz)^a)^{a^{-2}k\zeta}.$$

Taking $a \in K$ to be such that $a^2 = k$ and setting $\omega' = (\omega x)^a$ and $y = (xz)^a$, we see that $f(\omega') = (\omega' y)^\varsigma$. By (5), $y \neq 1$. $\qquad\square$

We will assume from here on in §2 that the elements ω_i have been chosen in such a way that $f(\omega_i) = (\omega_i y_i)^\varsigma$, $y_i \in Q_0^\#$.

In (10) to (18), we let ω denote one of the elements ω_i; we set $y_i = y = (0, \alpha)$.

(10) *Let* $a, b \in K$ *be such that* $b^{1+\theta} = \alpha + a^{-(1+\theta)}$. *Then*

$$f(\omega s^a) = (f(\omega s^b) s^a)^{\varsigma a^{-2}}.$$

Proof. By (2),

$$f(\omega s^a) = f(f(\omega) s^{a^{-1}})^{a^{-2}} s^{a^{-1}} = f(\omega y s^{a^{-1}})^{\varsigma a^{-2}} s^{a^{-1}} = (f(\omega s^b) s^a)^{\varsigma a^{-2}}. \qquad\square$$

We denote by τ the mapping inverse to the mapping $u \mapsto u^{1+\theta} : F^* \to F^*$ which is bijective since θ is of odd order. As $f(\omega) = (\omega y)^\varsigma$, (10) allows us to define inductively sequences (u_i), (v_i) and (d_i) such that u_i, $v_i \in F$, $d_i \in KW$ and $f(\omega(0, u_i)) = (\omega(0, v_i))^{d_i}$, namely,

(11) $u_1 = 0, \quad v_1 = \alpha, \quad d_1 = \varsigma,$

and, if $u_i \neq \alpha$,

$$u_{i+1} = \frac{1}{\alpha + u_i}, \quad v_{i+1} = v_i + u_{i+1} d_i^{-(1+\sigma)}, \quad d_{i+1} = d_i \varsigma u_{i+1}^{-2\tau}.$$

These sequences stop as soon as there is an index i such that $u_i = \alpha$. We set

$$\begin{pmatrix} a_i \\ b_i \end{pmatrix} = \begin{pmatrix} 0 & 1 \\ 1 & \alpha \end{pmatrix}^i \begin{pmatrix} 0 \\ 1 \end{pmatrix} \quad (i \geq 0, a_i, b_i \in F).$$

By induction on i, we see that, if $u_i \neq \alpha$, then $b_i \neq 0$ and $u_{i+1} = a_i/b_i$.

Let β and β^{-1} be the roots of the characteristic polynomial $X^2 + \alpha X + 1$ of the matrix

$$\begin{pmatrix} 0 & 1 \\ 1 & \alpha \end{pmatrix}.$$

Then

(12) $\beta + \beta^{-1} = \alpha$, $\beta \in E$ *and, if* $\beta \notin F$, *then* $\beta^{-1} = \beta^q$.

As $\alpha \neq 0$,

$$\begin{pmatrix} 0 & 1 \\ 1 & \alpha \end{pmatrix} = \begin{pmatrix} 1 & 1 \\ \beta & \beta^{-1} \end{pmatrix} \begin{pmatrix} \beta & 0 \\ 0 & \beta^{-1} \end{pmatrix} \begin{pmatrix} 1 & 1 \\ \beta & \beta^{-1} \end{pmatrix}^{-1}$$

whence

$$\begin{pmatrix} 0 & 1 \\ 1 & \alpha \end{pmatrix}^i = \frac{1}{\alpha} \begin{pmatrix} 1 & 1 \\ \beta & \beta^{-1} \end{pmatrix} \begin{pmatrix} \beta^i & 0 \\ 0 & \beta^{-i} \end{pmatrix} \begin{pmatrix} \beta^{-1} & 1 \\ \beta & 1 \end{pmatrix}$$

$$= \frac{1}{\alpha} \begin{pmatrix} \beta^{i-1} + \beta^{-i+1} & \beta^i + \beta^{-i} \\ \beta^i + \beta^{-i} & \beta^{i+1} + \beta^{-i-1} \end{pmatrix}.$$

It follows that $a_i = \dfrac{1}{\alpha}(\beta^i + \beta^{-i})$, $\quad b_i = \dfrac{1}{\alpha}(\beta^{i+1} + \beta^{-i-1})$ and

$$(13) \qquad\qquad u_i = \frac{\beta^{i-1} + \beta^{-i+1}}{\beta^i + \beta^{-i}}.$$

By (11), we then obtain, by induction on i, the fact that

$$(14) \qquad\qquad d_i = \zeta^i \left(\frac{\beta^i + \beta^{-i}}{\alpha} \right)^{2\tau}.$$

(15) *The sequences (u_i), (v_i) and (d_i) are defined up to and including $i = m - 1$, and $u_{m-1} = \alpha = \beta^{m-1} + \beta^{-m+1}$. Every $u \in F$ such that $\overline{f(\omega(0,u))}$ is in the orbit of $\overline{\omega}$ under KW is one of the elements u_i.*

Proof. Let m_1 be the last index for which u_i is defined. Then $m_1 \leq m - 1$ as, if not, $d_m \in K$ by (14) which contradicts (5). Also $u_{m_1} = \alpha$ and $f(\omega(0,\alpha)) = (\omega(0,v_{m_1}))^{d_{m_1}}$. But $f(\omega) = (\omega(0,\alpha))^\zeta$, whence $f(\omega(0,\alpha)) = \omega^{\zeta^{-1}}$. By (14), it follows that

$$d_{m_1} = \zeta^{m_1} \left(\frac{\beta^{m_1} + \beta^{-m_1}}{\alpha} \right)^{2\tau} = \zeta^{-1}.$$

Since $K \cap W = 1$, we then have $\beta^{m_1} + \beta^{-m_1} = \alpha$ and $\zeta^{m_1+1} = 1$. Since ζ is of order m, $m_1 = m - 1$. The elements u_i $(1 \leq i \leq m - 1)$ are pairwise distinct because the elements d_i are. The final assertion then follows from (8). $\qquad \square$

(16) *We have that β is a generator of W. In particular, $\beta^\sigma = \beta^{-1}$.*

Proof. For $1 \leq i \leq m - 1$, $b_{i-1} = \dfrac{1}{\alpha}(\beta^i + \beta^{-i}) \neq 0$ and so $\beta^i \neq 1$. By (15), β^{m-1} is a root of $X^2 + \alpha X + 1$ and so $\beta^{m-1} = \beta$ or β^{-1}, and, as $\beta^{m-2} \neq 1$, $\beta^m = 1$. $\qquad \square$

(17) *For $1 \leq i \leq m - 1$, $f(\omega(0,u_i)) = (\omega(0,u_i + \alpha))^{d_i}$.*

Proof. For $1 \leq i \leq m - 2$, $u_{i+1} \neq 0$ and, by (13) and (14),

$$
\begin{aligned}
\frac{u_i}{u_{i+1}} &= \frac{(\beta^{i-1} + \beta^{-i+1})(\beta^{i+1} + \beta^{-i-1})}{(\beta^i + \beta^{-i})^2} \\[2mm]
&= \frac{\beta^{2i} + \beta^{-2i} + \beta^2 + \beta^{-2}}{\beta^{2i} + \beta^{-2i}} \\[2mm]
&= 1 + \left(\frac{\beta + \beta^{-1}}{\beta^i + \beta^{-i}} \right)^2 = 1 + d_i^{-(1+\sigma)},
\end{aligned}
$$

and so $v_{i+1} + u_{i+1} = v_i + (1 + d_i^{-(1+\sigma)})u_{i+1} = v_i + u_i$, whence $v_i + u_i = v_1 + u_1 = \alpha$ for $1 \leq i \leq m - 1$. $\qquad \square$

(18) $(h(\omega)\zeta^{-1})^m = 1$.

Proof. By (13) and (16), $u_i^\theta = u_i$ and so $u_i^{2\tau} = u_i$. By (H6) and (1), we see that, for $2 \leq i \leq m - 1$,

$$
\begin{aligned}
h(\omega(0, u_i)) &= h(\omega)h((\omega(0, \alpha))^\zeta(0, u_i^{-1}))u_i^{2\tau} \\
&= h(\omega)h(\omega(0, u_{i-1}))^\zeta u_i \\
&= (h(\omega)\zeta^{-1})h(\omega(0, u_{i-1}))(\zeta u_i)
\end{aligned}
$$

so that, by induction,

$$
h(\omega(0, u_i)) = (h(\omega)\zeta^{-1})^i \zeta^i \left(\frac{\alpha}{\beta^i + \beta^{-i}} \right)
$$

for $1 \leq i \leq m - 1$. In particular, $h(\omega(0, \alpha)) = (h(\omega)\zeta^{-1})^{m-1}\zeta^{-1}$. But, by (H4),

$$
h(\omega(0, \alpha)) = h(f(\omega)^{\zeta^{-1}}) = \zeta h(\omega)^{-1}\zeta^{-1},
$$

whence $(h(\omega)\zeta^{-1})^m = 1$. $\qquad\square$

In (19) and (20), we assume that $n \geq 2$ and we set $y_1 = (0, \alpha_1)$ and $y_2 = (0, \alpha_2)$ so that $f(\omega_1) = (\omega_1(0, \alpha_1))^\zeta$ and $f(\omega_2) = (\omega_2(0, \alpha_2))^\zeta$. Let (u_i), (v_i) and (d_i) be the sequences defined in (11) using $\alpha = \alpha_1$, and let (u_i'), (v_i') and (d_i') be the analogous sequences defined using $\alpha = \alpha_2$. By (7) and (8), there are elements $x_1, x_2 \in F$ and $k \in K$ such that

$$
f(\omega_1(0, x_1)) = (\omega_2(0, x_2))^k.
$$

(19) *For $0 \leq i \leq m - 1$,*

(a) $\quad f\left(\omega_2\left(0, x_2 + \dfrac{1}{k^{1+\theta}(x_1 + u_i)}\right)\right) = \left(\omega_1\left(0, v_i + \dfrac{1}{d_i^{1+\sigma}(x_1 + u_i)}\right)\right)^{e_i},$

(b) $\quad f\left(\omega_1\left(0, x_1 + \dfrac{1}{k^{1+\theta}(x_2 + u_i')}\right)\right) = \left(\omega_2\left(0, v_i' + \dfrac{1}{d_i'^{1+\sigma}(x_2 + u_i')}\right)\right)^{e_i'},$

where $e_i = kd_i(x_1 + u_i)^{2\tau}$ and $e_i' = kd_i'(x_2 + u_i')^{2\tau}$.

Proof. Since $\overline{f(\omega_1(0, x_1))}$ is not in the orbit of $\overline{\omega_1}$ under KW, $x_1 \neq u_i$. Let $a \in K$ be such that $x_1 + a^{-(1+\theta)} = u_i$. Then, by (2),

$$
\begin{aligned}
f(\omega_2(0, x_2)s^{ak^{-1}})^{k^{-1}} &= f((\omega_2(0, x_2))^k s^a) &= f(\omega_1(0, x_1)s^{a^{-1}})^{a^{-2}}s^{a^{-1}} \\
&= f(\omega_1(0, u_i))^{a^{-2}}s^{a^{-1}} &= (\omega_1(0, v_i))^{d_i a^{-2}}s^{a^{-1}} \\
&= (\omega_1(0, v_i)s^{ad_i^{-1}})^{d_i a^{-2}},
\end{aligned}
$$

which proves (a). But

$$f(\omega_2(0, x_2))^{k^{-1}} = \omega_1(0, x_1)$$

so that $f(\omega_2(0, x_2)) = (\omega_1(0, x_1))^k$, and so we get (b) on interchanging the roles of ω_1 and ω_2. $\qquad\square$

(20) *We have $\alpha_1 = \alpha_2$; also, with $\alpha = \alpha_1$,*

$$f(\omega_1(0, x)) = (\omega_2(0, x + \alpha))^{d(x)},$$

with $d(x) \in KW$, for all elements x such that $\overline{f(\omega_1(0, x))}$ is in the orbit of $\overline{\omega_2}$ under KW.

Proof. The elements x_1 and $x_1 + \dfrac{1}{k^{1+\theta}(x_2 + u_i')}$ are pairwise distinct, whence, by (8) and (19), so are all the elements x such that $\overline{f(\omega_1(0, x))}$ is in the orbit of $\overline{\omega_2}$ under KW. By (19)(a),

$$f\left(\omega_1\left(0, v_i + \frac{1}{d_i^{1+\sigma}(x_1 + u_i)}\right)\right) = \left(\omega_2\left(0, x_2 + \frac{1}{k^{1+\theta}(x_1 + u_i)}\right)\right)^{e_i^{-t}}$$

and, by (19)(b),

$$f\left(\omega_1\left(0, x_1 + \frac{1}{k^{1+\theta}(x_2 + u_{m-i}')}\right)\right) = \left(\omega_2\left(0, v_{m-i}' + \frac{1}{d_{m-i}'^{1+\sigma}(x_2 + u_{m-i}')}\right)\right)^{e_{m-i}'}.$$

By (14) and (19), $e_i^{-t} \in e_{m-i}'K$, and so, by (7),

$$(*)\qquad\qquad v_i + \frac{1}{d_i^{1+\sigma}(x_1 + u_i)} = x_1 + \frac{1}{k^{1+\theta}(x_2 + u_{m-i}')}\,,$$

$$(**)\qquad\qquad v_i' + \frac{1}{d_i'^{1+\sigma}(x_2 + u_i')} = x_2 + \frac{1}{k^{1+\theta}(x_1 + u_{m-i})}\,,$$

$$(***)\qquad\qquad d_i^{-t}(x_1 + u_i)^{2\tau} = d_{m-i}'(x_2 + u_{m-i}')^{2\tau}$$

for $1 \le i \le m - 1$.

From $(***)$ for $i = 1$, and then for $i = m - 1$, we obtain

$$x_1 = x_2 + \alpha_2 \quad \text{and} \quad x_1 + \alpha_1 = x_2,$$

and so $\alpha_1 = \alpha_2 = x_1 + x_2$; this proves the first statement of (20) and the second statement for $x = x_1$. Since $\alpha_1 = \alpha_2$, we obtain $u_i' = u_i$, $v_i' = v_i$ and $d_i' = d_i$.

By (13) and (11), $u_{m-i} = 1/u_{i+1} = u_i + \alpha$ for $1 \leq i \leq m - 2$, and so $x_1 + u_{m-i} = x_1 + \alpha + u_i = x_2 + u_i$ for $1 \leq i \leq m - 1$, and $(**)$ can be expressed as

$$v_i + \frac{1}{d_i^{1+\sigma}(x_2 + u_i)} = x_2 + \frac{1}{k^{1+\theta}(x_2 + u_i)} = x_1 + \frac{1}{k^{1+\theta}(x_2 + u_i)} + \alpha.$$

By (19)(b),
$$f(\omega_1(0, x)) = (\omega_2(0, x + \alpha))^{d(x)},$$

with $d(x) \in KW$, for $x = x_1 + \dfrac{1}{k^{1+\theta}(x_2 + u_i)}.$ $\qquad\square$

Proposition. *Suppose that D acts without fixed points on $(Q/Q_0)^\#$. Then there exists an index i, $1 \leq i \leq n$, such that $f(\omega) = (\omega^{-1})^\varsigma$ and $h(\omega) \in W$ for $\omega = \omega_i$.*

Proof. By [H], Kapitel V, Satz 8.15, the Sylow subgroups of D are cyclic. Then (18) implies that, if ω is one of the elements ω_i, then $h(\omega) \in W$. In fact, if p is a prime number, then, if x is the p-component of $h(\omega)\varsigma^{-1}$ and P is a Sylow p-subgroup of D containing x, $x^{m_p} = 1$ and $|P \cap W| = m_p$ since $W \triangleleft D$, whence $x \in W$.

Set $\omega_1^2 = (0, r)$ and let i and k be the indices such that $\overline{f(\omega_1^{-1})}$ is in the orbit of $\overline{\omega_i}$ under KW and $\overline{f(\omega_1^{-1}(0, \alpha))}$ is in the orbit of $\overline{\omega_k}$ under KW. Then, by (17) and (20),

$$\begin{aligned} f((\omega_k(0, r))^{KW}) &= (\omega_1(0, \alpha + r))^{KW}, \\ (f \circ j)((\omega_1(0, \alpha + r))^{KW}) &= \omega_1^{KW}, \\ (f \circ j)(\omega_1^{KW}) &= (\omega_i(0, \alpha + r))^{KW}. \end{aligned}$$

Moreover, by (H4), $h(\omega_k^{-1}(0, r)) \in KW$. By (H5), it follows that

$$(\omega_k^{-1}(0, r))^{KW} = (\omega_i(0, \alpha + r))^{KW},$$

whence $i = k$ and $\omega_i^2 = (0, \alpha)$. Therefore,
$$f(\omega_i) = (\omega_i(0, \alpha))^\varsigma = (\omega_i^{-1})^\varsigma.$$

$$\qquad\square$$

§3. Determination of f

Proposition. *Suppose that there are elements $\omega \in Q - Q_0$ and $\zeta \in W^\#$ such that $f(\omega) = (\omega^{-1})^\varsigma$ and $h(\omega) \in W$. Then $\theta = 1$ and $f(\rho) = (\overline{\rho}/y, 1/y)$ for all $\rho = (\overline{\rho}, y) \in Q - Q_0$.*

The proof is given in a number of stages.

(1) *For $a \in K$, $f(\omega s^a)^{\zeta^{-1}a^2} s^a = f(\omega s^a)^{\zeta^{-2}} \omega^{\zeta^{-1}}$.*

Proof. First of all,

$$(f \circ j)^3(\omega^{-1}) = (f \circ j)^2(\omega^{-\zeta}) = (f \circ j)(\omega^{-\zeta^2}) = \omega^{-\zeta^3}.$$

By (H5) and the fact that $h(\omega^{-1}) = h(\omega)^{-t} \in W$, it follows that $h(\omega^{-1}) = \zeta^{-3}$. Also, $f(\omega^{-1}) = \omega^{\zeta^{-1}}$ and $g(\omega^{-1}) = \omega^{\zeta}$ whence, by (2) and (3) of §2,

$$f(\omega^{-1}s^{a^{-1}}) = f(\omega^{\zeta^{-1}}s^a)^{a^2} s^a = f(\omega^{\zeta}s^a)^{\zeta^{-3}} \omega^{\zeta^{-1}}.$$

$\square$

(2) *For $a \in K$,*

$$\overline{f(\omega s^a)} = \frac{\overline{\omega}}{a^2 + \zeta^{-1}}.$$

(The right-hand side is calculated in E.)

Proof. By (1),

$$\zeta^{-1}a^2\overline{f(\omega s^a)} = \zeta^{-2}\overline{f(\omega s^a)} + \zeta^{-1}\overline{\omega}$$

and so $(a^2 + \zeta^{-1})\overline{f(\omega s^a)} = \overline{\omega}$. $\square$

(3) *We have $\theta = 1$ and $\omega^2 = (0, \zeta + \zeta^{-1})$.*

Proof. Put $\omega^2 = (0, \alpha)$. By (2) of §2,

$$\overline{f(\omega s^a)} = \overline{f(\omega^{-\zeta}s^{a^{-1}})^{a^{-2}}} = \overline{\zeta a^{-2}f(\omega(0, \alpha + a^{-(1+\theta)}))}$$

for $a \in K$. Thus, by (2), if $a \neq \alpha^{-\tau}$ and $b^{1+\theta} = \alpha + a^{-(1+\theta)}$, then

$$\frac{1}{a^2 + \zeta^{-1}} = \frac{\zeta a^{-2}}{b^2 + \zeta^{-1}},$$

whence $b^2 = \zeta + \zeta^{-1} + a^{-2}$. It follows that

$$b^{2(1+\theta)} = \alpha^2 + (a^{-2})^{1+\theta} = (\zeta + \zeta^{-1} + a^{-2})^{1+\theta}.$$

As $\zeta \in W$, $(\zeta + \zeta^{-1})^{\theta} = \zeta^{\sigma} + \zeta^{-\sigma} = \zeta + \zeta^{-1}$. The equation above can then be written as

$$(*)\qquad \alpha^2 + \zeta^2 + \zeta^{-2} + (\zeta + \zeta^{-1})(a^{-2} + a^{-2\theta}) = 0.$$

Therefore, $c = X + X^{\theta}$ is independent of X for $X \in F - \{0, \alpha^{2\tau}\}$. If $\theta \neq 1$, then $|F| \geq 8$ since θ is of odd order, and so there are elements $X, Y \in F$ such that $\{X, Y, X + Y\} \cap \{0, \alpha^{2\tau}\} = \emptyset$ and $c = X + Y + (X + Y)^{\theta} = c + c = 0$. Thus $X^{\theta} = X$ for $X \in F - \{0, \alpha^{2\tau}\}$, whence $\theta = 1$. Equation $(*)$ then gives $\alpha = \zeta + \zeta^{-1}$. $\square$

(4) *For all $y \in E$ such that $y + y^q = \overline{\omega}^{1+q}$, $f(\overline{\omega}, y) = (\overline{\omega}/y, 1/y)$.*

Proof. Let $\omega = (\overline{\omega}, x)$. By (1) and (3),

$$f(\overline{\omega}, x + a)^{\zeta^{-1}a}(0, a) = f(\overline{\omega}, x + a)^{\zeta^{-2}}(\overline{\omega}, x)^{\zeta^{-1}}$$

for $a \in F - \{0\}$. By (2),

$$f(\overline{\omega}, x + a) = \left(\frac{\overline{\omega}}{a + \zeta^{-1}}, \gamma(a) \right)$$

where $\gamma(a) \in E$. It follows that

$$\left(\frac{\zeta^{-1} a \overline{\omega}}{a + \zeta^{-1}}, a^2 \gamma(a) + a \right)$$
$$= \left(\frac{\zeta^{-2} \overline{\omega}}{a + \zeta^{-1}} + \zeta^{-1} \overline{\omega}, \gamma(a) + x + \frac{\zeta^{-1} \overline{\omega}^{1+q}}{a + \zeta^{-1}} \right).$$

As $\overline{\omega}^{1+q} = \zeta + \zeta^{-1}$ by (3), equality between the second entries of these pairs gives

$$(**) \qquad\qquad (a^2 + 1)\gamma(a) = x + a + \frac{1 + \zeta^{-2}}{a + \zeta^{-1}}.$$

For $a = 1$, this becomes $0 = x + 1 + (1 + \zeta^{-1}) = x + \zeta^{-1}$, whence $x = \zeta^{-1}$. Equation $(**)$ can then be written as

$$(a^2 + 1)\gamma(a) = a + \zeta^{-1} + \frac{1 + \zeta^{-2}}{a + \zeta^{-1}} = \frac{a^2 + 1}{a + \zeta^{-1}}.$$

Thus, for $a \in F - \{0, 1\}$, $\gamma(a) = 1/(a + \zeta^{-1})$. This means that $f(\overline{\omega}, y) = (\overline{\omega}/y, 1/y)$ for $y \in E - \{\zeta^{-1}, \zeta^{-1} + 1\}$ such that $y + y^q = \overline{\omega}^{1+q}$. For $y = \zeta^{-1}$, $(\overline{\omega}, y) = \omega$ and again we obtain $f(\omega) = \omega^{-\zeta} = (\overline{\omega}, \zeta)^\zeta = (\overline{\omega}/y, 1/y)$. Since $f(\omega^{-1}) = \omega^{\zeta^{-1}}$, all of these results remain valid if we replace ω by ω^{-1} and ζ by ζ^{-1}, In particular, $f(\overline{\omega}, y) = (\overline{\omega}/y, 1/y)$ for all $y \in E - \{\zeta + 1\}$ such that $y + y^q = \overline{\omega}^{1+q}$; this completes the proof as $\zeta + 1 \neq \zeta^{-1} + 1$. $\qquad\square$

(5) *For all* $\rho = (\overline{\rho}, y) \in Q - Q_0$, $f(\rho) = (\overline{\rho}/y, 1/y)$.

Proof. By (H3), if $f(\overline{\rho}, y) = (\overline{\rho}/y, 1/y)$ and if $d \in KW$,

$$f(d\overline{\rho}, d^{1+q}y) = f(\overline{\rho}, y)^{d^t} = (\overline{\rho}/y, 1/y)^{d^{-q}} = \left(\frac{\overline{\rho}}{d^q y}, \frac{1}{d^{1+q}y} \right).$$

If $\overline{\rho}$ is in the orbit of $\overline{\omega}$ under KW, then (5) follows from (4). If not, possibly on replacing ρ by an element of ρQ_0, we may assume by (8) of §2 that $\overline{f(\rho)}$ is in the orbit of $\overline{\omega}$ under KW. Put $\rho = (\overline{\rho}, x)$ and $f(\rho) = (\overline{\omega}', x')$. Then $(\overline{\rho}, x) = f(\overline{\omega}', x') = (\overline{\omega}'/x', 1/x')$ and so $\overline{\omega}' = \overline{\rho}/x$ and $x' = 1/x$.

By (2) of §2,

$$\begin{aligned}
f(\overline{\rho}, x + a) &= f(\overline{\omega}', x' + a^{-1})^{a^{-1}}(0, a^{-1}) \\
&= \left(\frac{\overline{\omega}'}{x' + a^{-1}}, \frac{1}{x' + a^{-1}} \right)^{a^{-1}}(0, a^{-1}) \\
&= \left(\frac{\overline{\omega}'}{ax' + 1}, a^{-1}\left(\frac{1}{ax' + 1} + 1 \right) \right) \\
&= \left(\frac{\overline{\rho}}{x + a}, \frac{1}{x + a} \right)
\end{aligned}$$

for $a \in F - \{0\}$, which completes the proof. $\qquad\qquad\qquad\qquad\square$

Corollary 1. *Under the hypothesis of the proposition, $O^{2'}(G) \cong \mathrm{PSU}(3, q)$. In particular, if $V = W$, then G is isomorphic to $\mathrm{PSU}(3, q)$ or to $\mathrm{PGU}(3, q)$.*

Proof. If G satisfies the hypothesis of the proposition, Q and f are well defined by the specification of q. By the lemma of §1, $O^{2'}(G) = \langle Q^x \mid x \in G \rangle$ is then determined up to isomorphism.

Suppose that $V = W$. Then, by the proposition of §2, the hypothesis of the proposition is satisfied. In particular, $\mathrm{PGU}(3, q)$ satisfies this hypothesis and so $O^{2'}(G) \cong O^{2'}(\mathrm{PGU}(3, q)) = \mathrm{PSU}(3, q)$. Therefore, $(q + 1)/(q + 1, 3) \leq |W|$ and, if $|W| = (q + 1)/(q + 1, 3)$, then G is isomorphic to $\mathrm{PSU}(3, q)$. If not, $|W| = q + 1$, and G is isomorphic to $\mathrm{PGU}(3, q)$ by the lemma of §1. $\quad\square$

Corollary 2. *If $G \cong \mathrm{PSU}(3, q)$ and if $\zeta \in W^{\#}$, then there is an element $\omega \in Q - Q_0$ such that $f(\omega) = \omega^{-\zeta}$ and $h(\omega) = \zeta^3$.*

Proof. As $\zeta^{-1} + \zeta^{-q} \neq 0$, there is an element $\overline{\omega} \in E - \{0\}$ such that $\overline{\omega}^{1+q} = \zeta^{-1} + \zeta^{-q}$. Then $\omega = (\overline{\omega}, \zeta^{-1}) \in Q$, $\omega^{-1} = (\overline{\omega}, \zeta)$ and $f(\omega) = (\zeta\overline{\omega}, \zeta) = \omega^{-\zeta}$. Applying (H5), we thus see that $h(\omega) = \zeta^3$. $\qquad\qquad\qquad\square$

Remark. The treatment of case (10.2) of Chapter II can also be accomplished by use of the proposition of this section. In fact, in this case, by (11) of Chapter II, $h(\omega) = 1$ if $\omega \in C_Q(P)^{\#}$, and we see that $f(\omega) = \omega^{-\zeta}$ for some $\zeta \in C_W(P)^{\#}$ by use of (5) of §2 of this chapter and by use of the fact that $C_Q(P)$ is a quaternion group and $|C_W(P)| = 3$.

§4. The case $V \neq W$

By the proposition of §2 and Corollary 1 to the proposition of §3, to complete the proof of Theorem A, we may assume that D has a subgroup P of prime order p such that $C_{Q/Q_0}(P) \neq 1$. Since $C_Q(P) \neq 1$, P has three fixed points on Ω and so is conjugate in D to a subgroup of V. We may assume that $P \subset V$. Since W acts fixed-point-freely on Q/Q_0, $P \cap W = 1$.

(1) *Let $U = O^{2'}(C_G(P))$. Then $U/(P \cap U) \cong \mathrm{PSU}(3, \ell)$ with $q = \ell^p$ and $\ell > 2$.*

Proof. By (C1), $C_G(P)$ has 2-rank ≥ 2 and, by Chapter I, §3, Proposition 1(c), $U/Z(U) \cong \mathrm{PSU}(3, \ell)$ for some $\ell > 2$ since st has order 3 and $C_Q(P)$ has exponent 4. Now $|C_{Q_0}(P)| = \ell$ and so $q = \ell^p$ since P acts on Q_0 as a group of field automorphisms (Chapter I, §2, Proposition 3). As $Z(U) \subset C_V(C_{Q_0}(P))$, $Z(U) \subset PW$ by the theorem of Galois. Since $PZ(U)$ centralizes $C_Q(P) \not\subset Q_0$, $PZ(U) \cap W = 1$ and so $Z(U) \subset P$. $\qquad\qquad\qquad\square$

(2) *There are elements $\omega \in Q - Q_0$, $\zeta \in W^{\#}$ and $\eta \in P$ such that η centralizes ω and ζ, $f(\omega) = \omega^{-\zeta}$ and $h(\omega) = \zeta^3 \eta^{-1}$.*

Proof. By the structure of PSU$(3, \ell)$, $(V \cap U)/(P \cap U)$ centralizes $C_{Q_0}(P)$. Thus, by the theorem of Galois, $V \cap U \subset PW$ and, since $U \subset C_G(P)$, $V \cap U \subset P \times C_W(P)$. Furthermore, $|(V \cap U)/(P \cap U)| = (\ell + 1)/(\ell + 1, 3) \neq 1$ since $\ell > 2$. Let $\zeta_1 \in (V \cap U) - (P \cap U)$ and $\zeta \in C_W(P)$ be such that $\zeta_1 \in \zeta P$. If f_1 and h_1 denote the mappings f and h relative to U, $U \cap H$ and t, then, by Corollary 2 of the proposition of §3, there is an element $\omega \in (Q - Q_0) \cap U$ such that $f_1(\omega) \in \omega^{-\zeta_1}(P \cap U)$ and $h_1(\omega) \in \zeta_1^3(P \cap U)$. By the uniqueness of the canonical form of an element of $G - H$, $f(\omega) = f_1(\omega) = \omega^{-\zeta_1} = \omega^{-\zeta}$ and $h(\omega) = h_1(\omega) \in \zeta^3 P$. $\qquad\square$

In the following calculations, we again identify $Q \rtimes KW$ with the group $S_1 \rtimes K_1 W_1$ of Chapter III, §3. Then, by Proposition 2 of Appendix I, η acts as a semilinear mapping on $Q/Q_0 \cong E$. Let μ denote the automorphism of the field E associated with η. Thus, for $x \in E$, x^η is given by the action of η on Q/Q_0 and the isomorphism $Q/Q_0 \cong E$, while x^μ is given by the action of η on KW and the isomorphism $KW \cong K_1 W_1$.

Put $\omega^2 = (0, \alpha)$. Let $a \in K$ be such that $a \neq \alpha^{-\tau}$ and let $b \in K$ be such that $b^{1+\theta} = \alpha + a^{-(1+\theta)}$. Then, by (2) of §2,

(3) $\overline{f(\omega s^a)} = \overline{f(\omega^{-\zeta} s^{a^{-1}})^{a^{-2}}} = \zeta a^{-2} \overline{f(\omega s^b)}$.

(For $\rho \in Q$, $\bar{\rho}$ denotes the image of ρ in $E \cong Q/Q_0$).

By (2) and (3) of §2,

$$f(\omega^{-1} s^{a^{-1}}) = f(f(\omega^{-1})s^a)^{a^2} s^a = f(g(\omega^{-1})s^a)^{h(\omega^{-1})^t} f(\omega^{-1})$$

for $a \in K$, and so

$$f(\omega^{\zeta^{-1}} s^a)^{a^2} s^a = f(\omega^\zeta s^a)^{\zeta^{-3}\eta} \omega^{\zeta^{-1}},$$

whence

(4) $a^2 \overline{f(\omega s^a)} = \zeta^{-1} \overline{f(\omega s^a)}^{\eta} + \bar{\omega}$.

Assume again that $a \neq \alpha^{-\tau}$ and, in (4), replace $\overline{f(\omega s^a)}$ by the right-hand side of (3) to see that

(5) $\zeta \overline{f(\omega s^b)} = a^{-2\mu} \overline{f(\omega s^b)}^{\eta} + \bar{\omega}$.

Equation (4) holds with b in the place of a so that

(6) $b^2 \overline{f(\omega s^b)} = \zeta^{-1} \overline{f(\omega s^b)}^{\eta} + \bar{\omega}$.

Suitable linear combinations of (5) and (6) then show that $a^{2\mu} + b^2 \neq 0$ and that:

(7)
$$\overline{f(\omega s^b)} = \frac{a^{2\mu} + \zeta}{\zeta(a^{2\mu} + b^2)}\,\bar{\omega}.$$

(8)
$$\overline{f(\omega s^b)}^{\eta} = \frac{b^2 + \zeta}{b^2 a^{-2\mu} + 1}\,\bar{\omega}.$$

These equations imply that

$$(9) \qquad \frac{(\zeta^{-1} + a^{-2\mu})^\mu}{(1 + b^2 a^{-2\mu})^\mu} = \frac{b^2 + \zeta}{1 + b^2 a^{-2\mu}}.$$

Thus,

$$\lambda = \frac{\zeta^{-1} + a^{-2\mu^2}}{b^2 + \zeta} \in F$$

and $\lambda\zeta^2 + (\lambda b^2 + a^{-2\mu^2})\zeta + 1 = 0$. As $\zeta^{1+q} = 1$ and $\zeta \notin F$, it follows that $\lambda = 1$ and that $b^2 + a^{-2\mu^2} = \zeta + \zeta^{-1}$. The denominators on the two sides of (9) are then equal and $b^2 a^{-2\mu} = (\zeta + \zeta^{-1} + a^{-2\mu^2})a^{-2\mu}$ is fixed by μ. It follows that

$$(10) \quad (\zeta + \zeta^{-1} + X^\mu)X = (\zeta + \zeta^{-1} + X^{\mu^2})X^\mu$$

for $X \in F - \{0, \alpha^{2\tau}\}$.

Let $X \in F - \{0, \alpha^{2\tau}, \alpha^{2\tau} + 1\}$. Writing (10) with $X + 1$ in place of X and subtracting (10) from the result, we see that $X^{\mu^2} = X$. It follows that $\mu = 1$ since μ has odd order and, if $\mu \neq 1$, then $|F| \geq 8$. Thus $\eta \in W$ and $h(\omega) \in W$. We then conclude by using Corollary 1 of the proposition of §3. $\qquad\qquad\square$

Appendix I.
A Special Case of a Theorem of Huppert

Here we give a proof of a proposition which is a particular case of a theorem of Huppert on solvable doubly transitive permutation groups (see, for example, [HB], Chapter XII, § 7).

Proposition 1. *Let D be a group of odd order which acts faithfully on an elementary abelian q-group E (q prime) and which is transitive on $E^{\#}$. Then $F(D)$ is cyclic and acts without fixed points on E, and $D/F(D)$ is abelian.*

(Note that, under the hypotheses of this proposition, $E \rtimes D$ acts doubly transitively on E.)

Lemma. *Let p be a prime number, $p \neq 2$, and let P be a p-group acting faithfully on the elementary abelian q-group E. Assume that $|P_a|$ is independent of a for $a \in E^{\#}$. Then P is cyclic and acts without fixed points on E.*

Proof. We will denote the operation in E additively and consider E as an $\mathbf{F}_q[P]$-module.

(1) <u>Preliminary steps.</u>

Assume first of all that $E = E_1 \oplus \cdots \oplus E_r$ where $r \geq 2$ and the E_i are subspaces of E permuted by P (i.e., $(E_i)g$ is one of the subspaces E_j for $g \in P$ and $1 \leq i \leq r$).

Let $a \in E_1^{\#}$ and $b \in E_2^{\#}$. If $x \in P_{a+b}$, then $(a + b)x = ax + bx = a + b$. As P permutes the subspaces E_i, it follows that $ax = a$ and $bx = b$, or that $ax = b$ and $bx = a$. But the second case is impossible since x has odd order. Thus $P_{a+b} = P_a \cap P_b$. By hypothesis, $|P_{a+b}| = |P_a| = |P_b|$, and so $P_a = P_b$. As this is true for all $a \in E_1^{\#}$, P_a centralizes E_1. Since $P_a = P_b$ for all $b \in E_2^{\#}$, P_a centralizes E_2, and, in the same way, P_a centralizes E_i for $i > 1$. Therefore P_a centralizes E, whence $P_a = 1$. By the hypothesis, $P_a = 1$ for all $a \in E^{\#}$ and so P acts without fixed points on E. By [H], Kapitel V, Satz 8.15, it follows that P is cyclic.

(2) <u>Conclusion of the proof.</u>

By (1), we may assume that P acts irreducibly on E. Suppose that P is cyclic. Then, if $x \in P^{\#}$, $C_E(x)$ is a subgroup of E invariant under P and distinct from E so that $C_E(x) = 0$ and so P acts without fixed points on E. We may thus assume that P is not cyclic.

Then, by [H], Kapitel III, Hilfsatz 7.5, P contains a normal subgroup R of type (p, p). By Schur's Lemma ([Is], (1.5)), the ring of $\mathbf{F}_q[P]$-endomorphisms of E is a field. As $Z(P)$ is a subgroup of the multiplicative group of this finite field, $Z(P)$ is cyclic. Thus $|R \cap Z(P)| = p$ and P transitively permutes the set $\{T_i \mid i = 1, \ldots, p\}$ of subgroups of R of order p distinct from $R \cap Z(P)$. Since $C_E(R \cap Z(P))$ is a subspace of E invariant under P, it follows that

$C_E(R \cap Z(P)) = 0$. Let $E_i = C_E(T_i)$. We then know that $E = \sum E_i$ and that P acts on the set of subspaces E_i.

We show that the sum of the subspaces E_i is a direct sum. Assume that the sum $E_1 + \cdots + E_{k-1}$ is direct, and let $x \in E_k \cap (E_1 + \cdots + E_{k-1})$. Thus $x = x_1 + \cdots + x_{k-1}$ with $x_i \in E_i$. If $t \in T_k$, $xt = x_1 t + \cdots + x_{k-1} t = x_1 + \cdots + x_{k-1}$. As t acts on $E_i = C_E(T_i)$, it follows that t centralizes x_i for $i < k$ and so $R = \langle T_k, T_i \rangle$ centralizes x_i. Thus, $x_i = 0$ and so $x = 0$ and the sum $E_1 + \cdots + E_k$ is direct.

But then, by (1), P is cyclic contrary to assumption. ☐

Proof of Proposition 1. Let $F = F(D)$, let p be an odd prime number and let $P = O_p(F)$. Let $a, b \in E^\#$. Since $P \triangleleft D$ and D is transitive on $E^\#$, P_a and P_b are conjugate in D. Therefore, by the lemma, P is cyclic and acts without fixed points on E. Since $F = \prod_p O_p(F)$, it follows that F is cyclic and acts without fixed points on E.

By theorems of Feit-Thompson and of Fitting ([H], Kapitel III, Satz 4.2), $C_D(F) = F$, whence D/F is isomorphic to a subgroup of $\mathrm{Aut}(F)$ and, consequently, is abelian. ☐

Proposition 2. *Let U be a group acting faithfully on an elementary abelian group E of order p^n (p prime). Let T be a cyclic normal subgroup of U which acts irreducibly on E.*

(a) The subring $F = \mathbf{F}_p[T]$ of $\mathrm{End}(E)$ is a field with p^n elements and E is a vector space over F of dimension 1.

(b) U is a group of semilinear mappings from this vector space to itself. If $s \in E^\#$, then $C_U(s)$ is isomorphic to a group of field automorphisms of F.

Proof. We use additive notation for the operation in E and assume that U acts on E on the left.

(a) By Schur's Lemma, $F_1 = \mathrm{End}_T(E)$ is a finite field and, as T is commutative, $F = \mathbf{F}_p[T] \subset F_1$. Since F is a subring of a finite field, F is a field. But E is an irreducible F-module, and so an F-space of dimension 1, whence $|F| = |E|$.

(b) Let $u \in U$. Then u is an automorphism of E in its additive structure. Let $s \in E^\#$ and, for $\lambda \in F$, let $\sigma(\lambda)$ be that element of F for which $u(\lambda s) = \sigma(\lambda) u(s)$. Then $\sigma(\lambda + \mu) = \sigma(\lambda) + \sigma(\mu)$ for $\lambda, \mu \in F$. As U acts on $E \rtimes T$,

$$u(\lambda \mu s) = u \lambda u^{-1} \cdot u(\mu s) = u \lambda u^{-1} \cdot \sigma(\mu) u(s)$$

for $\lambda \in T$ and $\mu \in F$. In particular, for $\mu = 1$, $u(\lambda s) = u \lambda u^{-1} \cdot u(s)$. Thus $\sigma(\lambda \mu) = \sigma(\lambda) \sigma(\mu)$. As σ is additive, we see that $\sigma(\lambda \mu) = \sigma(\lambda) \sigma(\mu)$ for all elements $\lambda, \mu \in F$. Therefore σ is a field automorphism. As every $x \in E$ is of the form μs ($\mu \in F$), $u(\lambda x) = \sigma(\lambda) u(x)$ for all $\lambda \in F$ and all $x \in E$ whence u is semilinear. Furthermore, the mapping which associates to $u \in C_U(s)$ the automorphism σ such that $u(\lambda s) = \sigma(\lambda) s$, is an isomorphism of groups. ☐

Appendix II.
On Near-Fields

A finite set F, equipped with two operations $+$ and $\cdot$, is called a *near-field* if:

(1) F is a commutative group under the operation $+$ (the identity element is denoted by 0).

(2) $F - \{0\}$ is a group under the operation $\cdot$ (the identity element is denoted by 1 and juxtaposition is often used to denote this operation).

(3) The law $(a + b)c = ac + bc$ holds for all $a, b, c \in F$.

In [Z], Zassenhaus classified finite near-fields but here we need only certain elementary results.

If F is a near-field, we set $\mathcal{L}(F) = F \rtimes F^*$, where F^* is the multiplicative group $F - \{0\}$ acting on F by multiplication on the right. Then F^* acts regularly on $F - \{0\}$. Conversely, if a group M acts on an abelian group F and M is regular on $F^{\#}$, we can equip F with the structure of a near-field such that $\mathcal{L}(F) \cong F \rtimes M$: it suffices to choose an element 1 in $F - \{0\}$ and to put $(1x)(1y) = 1(xy)$ for $x, y \in M$.

Let F be a near-field. As F^* acts transitively on $F - \{0\}$, F is an elementary abelian f-group for some prime number f, which is called the *characteristic* of F.

Proposition 1. *Suppose that G satisfies* (A1) *and* (A2) *and that G has 2-rank 1. Then there is a near-field F, a group Σ of automorphisms of F and an isomorphism from G to $\mathcal{L}(F) \rtimes \Sigma = (F \rtimes F^*) \rtimes \Sigma$ which identifies Q with F^* and D with Σ. Moreover, H has only one involution and, if u and v are distinct involutions of G, then the order of uv is equal to the characteristic of F.*

Proof. Let u be an involution of H. Since G has 2-rank 1, a Sylow 2-subgroup of G is cyclic or generalized quaternion ([H], Kapitel III, Satz 8.2). By the Brauer-Suzuki Theorem (see [S2], Chapter 6, § 2.2, Example 3), $G = O_{2'}(G)C_G(u)$. By the Feit-Thompson Theorem, $O_{2'}(G)$ is solvable. By [H], Kapitel II, Satz 3.2, there is an elementary abelian normal subgroup F of G such that $G = F \rtimes H$, and F acts regularly on Ω. Since Q acts regularly on $\Omega - \{H\}$, Q acts regularly on $F^{\#}$. We can then equip F with the structure of a near-field such that $F \rtimes Q \cong F \rtimes F^*$. The group D fixes an element of $F^{\#}$ (the element x such that $H^x = H^t$) which we can take as the multiplicative identity element 1 of the near-field. Moreover, D acts faithfully on F since it acts faithfully on Ω, and, as D acts on $F \rtimes Q$, D can thus be identified with a group of automorphisms of the near-field F. Since Q acts regularly on $F^{\#}$, $C_F(u) = 1$ and the elements of F are inverted by u (Chapter I, § 1, Lemma). It follows that H has a unique involution. Let u and v be distinct involutions of G. Since G/F has a unique involution, $uv \in F^{\#}$. $\qquad\square$

The near-field $F_{r^2,2}$

Let r be a power of an odd prime number and let $K = \mathbf{F}_{r^2}$. For $x, y \in K$, we set

$$x \circ y = \begin{cases} xy & \text{if } y \text{ is a square in the field } K, \\ x^r y & \text{if not.} \end{cases}$$

We can then check that K, equipped with the operations $+$ and $\circ$, is a near-field. This near-field is denoted by $F_{r^2,2}$.

Proposition 2. *Let F be a finite near-field whose multiplicative group has a cyclic subgroup of index 2. Then either F is a field or there is an integer r which is a power of an odd prime number, for which F is isomorphic to $F_{r^2,2}$. In the second case, $|Z(F^*)| = r - 1$.*

Proof. Let A be the cyclic subgroup of index 2 in F^*. Suppose that A does not act irreducibly on the additive group F. Then $F = F_1 \oplus F_2$, $F_i \neq 0$ for $i = 1, 2$, and A acts without fixed points on F_i, whence

$$|F_i| \geq |A| + 1 \quad \text{and} \quad |F| = 2|A| + 1 \geq (|A| + 1)^2,$$

which is impossible. Thus A acts irreducibly on F.

We denote the multiplication in the near-field F by $\circ$. By Appendix I, Proposition 2, we can define a multiplication $\cdot$ on F which makes it a field K which has the same addition and the same multiplicative identity 1 as F and is such that, for $y \in A$ and $x \in F$, $x \circ y = xy$. Moreover, for $y \in F - \{0\}$, the mapping $x \mapsto x \circ y$ is semilinear and so there is an automorphism σ_y of K such that

$$x \circ y = (1x) \circ y = x^{\sigma_y}(1 \circ y) = x^{\sigma_y} y.$$

We see that $y \mapsto \sigma_y$ is a homomorphism from F^* to $\mathrm{Aut}(K)$ whose kernel contains A. If this kernel is F^*, then $F^* \cong K^*$ and F is a field. Otherwise, for $y \in F^* - A$, σ_y is of order 2 and so there is an integer r such that $K = \mathbf{F}_{r^2}$ and $x^{\sigma_y} = x^r$. As A is also the subgroup of index 2 in K^* and $x \circ y = x^{\sigma_y} y$ for $x \in F$ and $y \in F^*$, we see that $F \cong F_{r^2,2}$.

In this case, $Z(F^*) \subset A$ as otherwise F^* would be abelian and F would be a field. If $x \in A$ and $y \in F^* - A$, then $x \circ y = x^r y$ and $y \circ x = yx$; it follows that $x \in Z(F^*)$ if and only if $x^r = x$, i.e., $x \in \mathbf{F}_r^*$. $\qquad\square$

Appendix III.
On Suzuki 2-Groups

Let V and W be vector spaces of finite dimension over $\mathbf{F}_2$. Recall that, in characteristic 2, a mapping $q : V \to W$ is *quadratic* if the mapping

$$(x, y) \mapsto q(x + y) + q(x) + q(y)$$

is bilinear.

Lemma 1. (a) *Let P be a 2-group and W a subgroup of $Z(P)$ such that W and $V = P/W$ are elementary abelian. The mapping $x \mapsto x^2$ induces a quadratic mapping from V to W, where V and W are considered as vector spaces over $\mathbf{F}_2$.*

(b) *Let V and W be vector spaces of finite dimension over $\mathbf{F}_2$. For every quadratic mapping $q : V \to W$, there is a central extension $W \overset{\iota}{\to} P \overset{\pi}{\to} V$ such that the map from $P/\iota(W)$ to $\iota(W)$ defined in (a) is identified with q.*

(c) *Let V, W, V' and W' be vector spaces of finite dimension over $\mathbf{F}_2$, let $W \to P \overset{\pi}{\to} V$ and $W' \to P' \overset{\pi'}{\to} V'$ be central extensions with associated quadratic mappings $q : V \to W$ and $q' : V' \to W'$. Let $f : V \to V'$ and $g : W \to W'$ be isomorphisms. For there to be an isomorphism $P \to P'$ which induces f on V and g on W, it is necessary and sufficient that $g \circ q = q' \circ f$.*

(d) *If V and W are vector spaces of finite dimension over $\mathbf{F}_2$ and if*

$$W \to P \to V$$

is a central extension, then the set of automorphisms of P which induce the identity on V and on W, is a group isomorphic to the additive group $\mathrm{Hom}(V, W)$.

Proof. (a) It is clear that $x \mapsto x^2$ induces a mapping q from V to W and that the commutator mapping induces a mapping $(x, y) \mapsto [x, y]$ from $V \times V$ to W. The latter is bilinear by [H], Kapitel III, Hilfsatz 1.2. If $\bar{x}$, $\bar{y}$ are the images in V of $x, y \in P$, then

$$q(\bar{x} + \bar{y}) = (xy)^2 = x^2 y^2 [y, x] = q(\bar{x}) + q(\bar{y}) + [y, x].$$

Thus q is a quadratic mapping.

(b) Let $(e_1, \ldots, e_n)$ be a basis for V over $\mathbf{F}_2$. Let $b : V \times V \to W$ be the bilinear mapping satisfying

$$b(e_i, e_j) = \begin{cases} q(e_i) & \text{if } i = j, \\ q(e_i + e_j) + q(e_i) + q(e_j) & \text{if } i < j, \\ 0 & \text{if } i > j. \end{cases}$$

Then $b(v, v) = q(v)$ for all $v \in V$ and it is sufficient to take as P the set $V \times W$ equipped with the operation

$$(v_1, w_1)(v_2, w_2) = (v_1 + v_2, b(v_1, v_2) + w_1 + w_2)$$

and to set $\iota(w) = (0, w)$ and $\pi(v, w) = v$ for all $v \in V$ and $w \in W$.

(c) Necessity is immediate. Suppose that $g \circ q = q' \circ f$. Let $(e_1, \ldots, e_n)$ be a basis for V; let $x_1, \ldots, x_n \in P$ be such that $\pi(x_i) = e_i$ and let $y_1, \ldots, y_n \in P'$ be such that $\pi'(y_i) = f(e_i)$. We identify W and W' with their images in P and P' respectively. By hypothesis, $g(x_i^2) = y_i^2$ and $g([x_i, x_j]) = [y_i, y_j]$ since $[x_i, x_j] = (x_i x_j)^2 x_i^2 x_j^2$. Each element x of P can be written in a unique way in the form

$$x = x_1^{a_1} \ldots x_n^{a_n} w \quad (0 \le a_i \le 1, w \in W).$$

Therefore, setting $h(x) = y_1^{a_1} \ldots y_n^{a_n} g(w)$, we see that h is an isomorphism from P to P' which induces f on V and g on W.

(d) By (c), every central extension of W by V is equivalent to one of the extensions constructed in (b). It can be checked that the set of those automorphisms of the group P constructed in (b) which induce the identity on V and on W, is the set of mappings $(v, w) \mapsto (v, h(v) + w)$ where h is a homomorphism $V \to W$. $\square$

Lemma 2. *Let F be a finite field of characteristic 2.*

(a) *The set of $\mathbf{F}_2$-linear mappings $F \to F$ is a vector space over F having as basis the set of automorphisms of the field F.*

(b) *The set of $\mathbf{F}_2$-bilinear mappings $F \times F \to F$ is a vector space over F having as basis the family of mappings $(x, y) \mapsto \sigma(x)\tau(y)$ where (σ, τ) runs through all pairs of automorphisms of F.*

(c) *The set of $\mathbf{F}_2$-quadratic mappings $F \to F$ is a vector space over F having as basis the family of mappings $x \mapsto \sigma(x)\tau(x)$ where $\{\sigma, \tau\}$ runs through all subsets of $\mathrm{Aut}(F)$ of size 1 or 2.*

Proof. (a) It is known that the automorphisms of the field F are linearly independent over F. For $|F| = 2^n$, $|\mathrm{Hom}_{\mathbf{F}_2}(F, F)| = |F|^n$ and $|\mathrm{Aut}(F)| = n$; the result follows.

(b) Suppose that, for some $\lambda_{\sigma\tau} \in F$, $\sum \lambda_{\sigma\tau}\sigma(x)\tau(y) = 0$ for all $x, y \in F$, where the sum is taken over all $(\sigma, \tau) \in \mathrm{Aut}(F) \times \mathrm{Aut}(F)$. Applying (a) twice, we then see that $\lambda_{\sigma\tau} = 0$ for all σ and for all τ. Thus the mappings $(x, y) \mapsto \sigma(x)\tau(y)$ are linearly independent over F. As the space of $\mathbf{F}_2$-bilinear mappings $F \times F \to F$ is of dimension n^2 over F, the result follows.

(c) Let $q(x) = \sum \lambda_{\sigma\tau}\sigma(x)\tau(x)$, where the sum is taken over all subsets $\{\sigma, \tau\} \subset \mathrm{Aut}(F)$ and $\lambda_{\sigma\tau} \in F$. Suppose that $q(x) = 0$ for all $x \in F$. Then

$$q(x + y) + q(x) + q(y) = \sum_{\{\sigma, \tau\}} \lambda_{\sigma\tau}(\sigma(x)\tau(y) + \sigma(y)\tau(x)) = 0.$$

It follows from (b) that $\lambda_{\sigma\tau} = 0$ for $\sigma \ne \tau$. Therefore,

$$q(x) = \sum_{\sigma} \lambda_{\sigma\sigma}\sigma(x)^2 = \sum_{\sigma} \lambda_{\sigma\sigma}\sigma(x^2) = 0$$

for all x, and so $\lambda_{\sigma\sigma} = 0$ by (a). The mappings $x \mapsto \sigma(x)\tau(x)$ are thus linearly independent over F. On the other hand, if q is a quadratic mapping $F \to F$, there is an $\mathbf{F}_2$-bilinear mapping $f : F \times F \to F$ such that $q(x) = f(x, x)$ (see Lemma 1(b)). By (b), q is thus an F-linear combination of mappings $x \mapsto \sigma(x)\tau(x)$. $\qquad\square$

Definition 1. A *Suzuki 2-group* is a 2-group P such that P is non-abelian, P has at least two involutions and there is a cyclic group K which acts faithfully on P and regularly on the set of involutions of P.

Definition 2. Let $F = \mathbf{F}_{2^n}$ and let φ be an automorphism of F. The group $A(n, \varphi)$ is defined as the central extension of F by F associated, in the sense of Lemma 1(b), with the quadratic mapping $x \mapsto x\varphi(x)$ from F to F. A 2-group is of *type A* if it is isomorphic to a group $A(n, \varphi)$ for some automorphism φ of $F = \mathbf{F}_{2^n}$ which is non-trivial and of odd order.

Definition 3. Let $F = \mathbf{F}_{2^n}$, let φ be an automorphism of F of odd order and let $\varepsilon \in F^*$ be such that $a\varphi(a) + \varepsilon a\varphi(b) + b\varphi(b) \neq 0$ for $a, b \in F^*$. The group $B(n, \varphi, \varepsilon)$ is defined as the central extension of F by $F \times F$ associated, in the sense of Lemma 1(b), with the quadratic mapping

$$(a, b) \mapsto a\varphi(a) + \varepsilon a\varphi(b) + b\varphi(b)$$

from $F \times F$ to F. A 2-group is of *type B* if it is isomorphic to a group $B(n, \varphi, \varepsilon)$.

We take the following theorem as given; it is proved in [Hi] (see also [HB], Chapter VIII, Theorem 7.9).

Theorem. *Let P be a Suzuki 2-group, $Z = Z(P)$, $q = |Z|$ and $F = \mathbf{F}_q$. Let K be as in Definition 1.*

(a) The set of involutions of P is $Z^{\#}$, and Z is an elementary abelian group.

(b) P/Z is an elementary abelian group of order q or q^2. Thus, $|P| = q^2$ or q^3 and P is of exponent 4.

(c) If $|P| = q^2$, then P is of type A. We can identify K with F^* in such a way that the actions of K on P/Z and on Z, identified with F, are given by $a^x = xa$ and $b^x = x\varphi(x)b$, where $x \in F^*$, $a \in F \cong P/Z$, $b \in F \cong Z$ and φ is as in Definition 2.

(d) If $|P| = q^3$, then P/Z is a direct sum of two $\mathbf{F}_2[K]$-modules which have order q.

(e) P is of type B if and only if P/Z is a direct sum of two isomorphic $\mathbf{F}_2[K]$-modules. Suppose that P is of type B and that φ is as in Definition 3. We can identify K with F^* in such a way that the actions of K on P/Z and on Z, identified with $F \times F$ and with F respectively, are given by $(a, b)^x = (xa, xb)$ and $c^x = x\varphi(x)c$, where $x \in F^*$, $(a, b) \in F \times F$ and $c \in F$.

Suzuki 2-groups which are not of type A or of type B are called of type C or of type D in [Hi]. We have no need of the explicit definition of type C or of type D.

Proposition 1. *Let $F = \mathbf{F}_{2^n}$, $E = F \times F$ and let $q : E \to F$ be the quadratic mapping associated with the group $B(n, 1, \varepsilon)$. There is a field structure on E, compatible with its structure of vector space over F, such that $q(x) = x\overline{x}$ for $x \in E$, where $x \mapsto \overline{x}$ is the automorphism of this field structure which is of order 2.*

Proof. Since $q(a, b) = a^2 + \varepsilon ab + b^2 \neq 0$ for $(a, b) \neq (0, 0)$, the polynomial $X^2 + \varepsilon X + 1$ is irreducible over F. Let $\alpha \in \mathbf{F}_{2^{2n}}$ be such that $\alpha^2 + \varepsilon\alpha + 1 = 0$. We identify $(a, b) \in E$ with $a + b\alpha$. As $\alpha + \overline{\alpha} = \varepsilon$ and $\alpha\overline{\alpha} = 1$, we then have $(a + b\alpha)\overline{(a + b\alpha)} = q(a + b\alpha)$ for $a, b \in F$. $\square$

According to this proposition, the structure of $B(n, 1, \varepsilon)$ is independent of ε if ε satisfies the condition of Definition 3. Put $B(n, 1) = B(n, 1, \varepsilon)$. If $P = B(n, 1)$, $F = \mathbf{F}_{2^n}$ and $E = \mathbf{F}_{2^{2n}}$, we then obtain a central extension $F \xrightarrow{\iota} P \to E$. Let $\alpha \in \operatorname{Aut}(P)$. One can check that $\iota(F) = \Omega_1(P)$; it follows that α induces isomorphisms f_α and g_α of additive groups such that the diagram

$$
\begin{array}{ccccc}
F & \longrightarrow & P & \longrightarrow & E \\
\downarrow g_\alpha & & \downarrow \alpha & & \downarrow f_\alpha \\
F & \longrightarrow & P & \longrightarrow & E
\end{array}
$$

is commutative.

Proposition 2. *In the above notation, the map $\alpha \mapsto f_\alpha$ is a surjective homomorphism from $\operatorname{Aut}(B(n, 1))$ onto the group of mappings $x \mapsto \lambda\sigma(x)$ from E to itself $(\lambda \in E^*, \sigma \in \operatorname{Aut}(E))$. The kernel of $\alpha \mapsto f_\alpha$ is an elementary abelian 2-group.*

Proof. Let $f : E \to E$ and $g : F \to F$ be the additive isomorphisms induced by a given automorphism of $B(n, 1)$. Then $g \circ q = q \circ f$ where $q(x) = x\overline{x}$ for $x \in E$. By Lemma 2(a), we can write

$$
f(x) = \sum_{i=0}^{2n-1} \lambda_i x^{2^i} \ (\lambda_i \in E) \ \ \text{and} \ \ g(x) = \sum_{i=0}^{n-1} \mu_i x^{2^i} \ (\mu_i \in F).
$$

Thus, for $x \in E$,

$$
(1) \qquad\qquad g(q(x)) = \sum_{i=0}^{n-1} \mu_i x^{2^i + 2^{i+n}};
$$

$$
(2) \qquad\qquad q(f(x)) = \left(\sum_{i=0}^{2n-1} \lambda_i x^{2^i} \right) \left(\sum_{i=0}^{2n-1} \lambda_i^{2^n} x^{2^{i+n}} \right).
$$

If i and j are integers modulo $2n$, then, by Lemma 2(c), the coefficients of $x^{2^i + 2^j}$ in (1) and (2) are equal since $g \circ q$ and $q \circ f$ can be considered as quadratic mappings $E \to E$. Thus,

(3) $$\lambda_i \lambda_{i-n}^{2^n} = 0 \ \text{ for all } \ i,$$

(4) $$\lambda_i \lambda_{j-n}^{2^n} + \lambda_j \lambda_{i-n}^{2^n} = 0 \ \text{ if } \ i - j \not\equiv 0, \ n \ \pmod{2n}.$$

(Indices are to be interpreted modulo $2n$.) Since $f \neq 0$, there is an index i for which $\lambda_i \neq 0$. Then, by (3) and (4),

$$\lambda_{i-n} = 0 \ \text{ and } \ \lambda_{j-n} = 0 \ \text{ if } \ j \not\equiv i, \ i - n \ \pmod{2n}.$$

Consequently, $\lambda_j = 0$ for $j \not\equiv i \pmod{2n}$.

Therefore, there is an element $\lambda \in E^*$ and an integer i such that $f(x) = \lambda x^{2^i}$. On comparing the coefficients of $x^{2^j + 2^{j+n}}$ in (1) and (2), we see that:

if $0 \leq i \leq n - 1$, $\mu_i = \lambda \overline{\lambda}$ and $\mu_j = 0$ for $j \neq i$;

if $n \leq i \leq 2n - 1$, $\mu_{i-n} = \lambda \overline{\lambda}$ and $\mu_j = 0$ for $j \neq i - n$.

The pairs (f, g) induced by the automorphisms of $B(n, 1)$ are thus of the form

$$f(x) = \lambda x^{2^i}, \ g(x) = \lambda \overline{\lambda} x^{2^i}, \ \lambda \in E^*, \ 0 \leq i \leq 2n - 1.$$

Conversely, if f and g are defined by these formulae, they are additive isomorphisms such that $g \circ q = q \circ f$, and so come from an automorphism of $B(n, 1)$. This completes the proof of the first part of the proposition, and the second part follows from Lemma 1(d). $\qquad\square$

Appendix IV.
The Feit-Sibley Theorem

In this appendix we prove a version of the Feit-Sibley Theorem adapted to our requirements. This theorem has its origin in the work of Feit, in particular, in [F]. In this paper, Feit proves a coherence theorem; it was generalized by Sibley in [Si].

Let G be a finite group, H a subgroup of G, $\mathcal{S}$ a subset of $\mathrm{Irr}(H)$ and τ a linear isometry from $\mathbf{Z}[\mathcal{S}]^{\circ}$ to $\mathbf{Z}[\mathrm{Irr}(G)]^{\circ}$. Recall that $(\mathcal{S}, \tau)$, or simply $\mathcal{S}$ if there is no ambiguity concerning τ, is called *coherent* if $|\mathcal{S}| \geq 2$ and if τ can be extended to a linear isometry from $\mathbf{Z}[\mathcal{S}]$ to $\mathbf{Z}[\mathrm{Irr}(G)]$.

The Feit-Sibley Theorem shows that $\mathcal{S}$ is coherent under certain special conditions on H and $\mathcal{S}$; this allows information on the characters of G to be drawn from the characters of H. First of all:

Lemma 1. (a) *Suppose that $\mathcal{S} = \mathcal{S}_0 \cup \{\psi\}$, that $(\mathcal{S}_0, \tau)$ is coherent and that there is a character $\chi_0 \in \mathcal{S}_0$ such that*

$$\chi_0(1) \mid \psi(1) \quad and \quad 2\chi_0(1)\psi(1) < \sum_{\chi \in \mathcal{S}_0} \chi(1)^2.$$

Then $(\mathcal{S}, \tau)$ is coherent.

(b) *If $|\mathcal{S}| \geq 2$ and if all $\chi \in \mathcal{S}$ have the same degree, then $(\mathcal{S}, \tau)$ is coherent.*

Proof. See [Is], Theorem 7.14 and Corollary 7.15. $\square$

Hypotheses and Notation. *G is a finite group and $H = Q \rtimes D$ is a proper subgroup of G. We assume that $(|D|, |Q|) = 1$ and that $Q \cap Q^x = 1$ for $x \in G - H$.*

(These hypotheses imply that Q is a Hall subgroup of G since, if P is a Sylow subgroup of Q and $P \neq 1$, then $N_G(P) \subset H$.)

$Q = S \times Q_1$, $|Q_1|$ and $|S|$ are relatively prime, D acts without fixed points on Q_1, Q_1 is not a 2-group and S is nilpotent.

$\mathcal{S} = \{\chi \in \mathrm{Irr}(H) \mid Q_1 \not\subset \mathrm{Ker}\,\chi\}$.

Set $S' = [S, S]$, $Q' = [Q, Q]$ and $d = |D|$. For $R \subset Q$, set

$$\mathcal{S}(R) = \{\chi \in \mathcal{S} \mid R \subset \mathrm{Ker}\,\chi\}.$$

Lemma 2. (a) *$\mathcal{S}$ is the set of characters $\mathrm{Ind}_Q^H \varphi$ induced from characters $\varphi \in \mathrm{Irr}(Q)$ for which $Q_1 \not\subset \mathrm{Ker}\,\varphi$.*

(b) *The mapping $\psi \mapsto \mathrm{Ind}_H^G \psi$ is an isometry from $\mathbf{Z}[\mathcal{S}]^{\circ}$ to $\mathbf{Z}[\mathrm{Irr}(G)]^{\circ}$.*

(c) *If d is odd and $\chi \in \mathcal{S}$, then $\overline{\chi} \neq \chi$.*

Proof. (a) Let $\varphi = \lambda\theta$, where $\lambda \in \mathrm{Irr}(S)$ and $\theta \in \mathrm{Irr}(Q_1) - \{1_{Q_1}\}$. If $x \in H$ is such that $\varphi^x = \varphi$, then $\theta^x = \theta$, whence $x \in Q$ since $Q_1 D$ is a Frobenius group ([Is], Theorem 6.34). Thus the inertia group of φ in H is Q and so $\mathrm{Ind}_Q^H \varphi$ is irreducible ([Is], Theorem 6.11). The restriction of $\mathrm{Ind}_Q^H \varphi$ to $Q_1 D$ is $\mathrm{Ind}_{Q_1}^{Q_1 D} \theta$. This restriction is therefore irreducible and $Q_1 \not\subset \mathrm{Ker}\, \mathrm{Ind}_Q^H \varphi$.

If $\chi \in \mathcal{S}$, $\mathrm{Res}_Q^H \chi$ has a constituent φ such that $Q_1 \not\subset \mathrm{Ker}\,\varphi$. Thus χ is a constituent of $\mathrm{Ind}_Q^H \varphi$ and so $\chi = \mathrm{Ind}_Q^H \varphi$.

(b) This follows from [Is], Lemma 7.7, since the elements of $\mathcal{S}$ vanish on $H - Q$ by (a).

(c) We have seen in (a) that the restriction of χ to the odd order group $Q_1 D$ is irreducible and non-principal, whence $\overline{\chi} \neq \chi$ by [H], Kapitel V, Satz 13.8. $\qquad\square$

Theorem. *If d is odd, then $\mathcal{S}$ is coherent with respect to the isometry of Lemma 2(b).*

We remark that $|\mathcal{S}(Q')| \geq 2$ since $O_{2'}(Q_1) \rtimes D$ is a Frobenius group of odd order. By Lemma 1(b), $\mathcal{S}(Q')$ is coherent.

(1) *Suppose that $|Q_1|$ is divisible by two prime numbers. Then $\mathcal{S}(S')$ is coherent.*

Proof. We may assume that Q_1 is not abelian. Let $Q_2 \subset [Q_1, Q_1]$ be such that $Q_2 \triangleleft H$ and $\mathcal{S}(S'Q_2)$ is coherent. Let $Q_3 \triangleleft Q_2$ be such that Q_2/Q_3 is a chief factor of H. It suffices to show that $\mathcal{S}(S'Q_3)$ is coherent because, if we take Q_2 minimal with respect to the conditions above, then $Q_2 = 1$. Suppose that $\mathcal{S}(S'Q_3)$ is not coherent. By Lemma 1(a), there is a character $\psi \in \mathcal{S}(S'Q_3)$ such that

$$\sum_{\chi \in \mathcal{S}(S'Q_2)} \chi(1)^2 \leq 2d\psi(1).$$

But $\sum_{\chi \in \mathcal{S}(S'Q_2)} \chi(1)^2 = |H/S'Q_2| - |H/S'Q_1| = d|S/S'|(|Q_1/Q_2| - 1)$, whence

$$(1.1) \qquad\qquad |Q_1/Q_2| - 1 \leq 2\psi(1).$$

Let $Z/Q_3 = Z(Q_1/Q_3)$. Then $\psi(1) = d\varphi(1)$, where $\varphi \in \mathrm{Irr}(Q/S'Q_3)$, and, by [Is], Corollary 2.30, $\varphi(1)^2 \leq |Q/SZ|$, whence

$$(1.2) \qquad\qquad \psi(1)^2 \leq d^2 |Q/SZ| = d^2 |Q_1/Z|.$$

Since Q_1 is nilpotent, $(Q_2/Q_3) \cap (Z/Q_3) \neq 1$, and, as Q_2/Q_3 is a chief factor of H, $Q_2 \subset Z$. Moreover, since $|Q_1|$ has two prime divisors, $Q_2 \subsetneq Z$. By (1.1) and (1.2),

$$(|Q_1/Q_2| - 1)^2 \leq 4d^2 |Q_1/Z|,$$

whence

$$|Q_1/Q_2|(|Q_1/Q_2| - 2) < 4d^2 |Q_1/Z|,$$

or

$$|Z/Q_2|(|Q_1/Q_2| - 2) < 4d^2.$$

But, since D acts on Q_1 without fixed points and $Q_2 \subsetneqq Z \subsetneqq Q_1$, $|Z/Q_2| \geq d+1$ and $|Q_1/Q_2| \geq (d+1)^2$, and so $(d+1)^2 - 2 < 4d$, whence $d \leq 2$, which is a contradiction. $\qquad\qquad\square$

(2) *Suppose that $\mathcal{S}(S')$ is coherent. Then $\mathcal{S}$ is coherent.*

Proof. We may assume that $S' \neq 1$. As in (1), we take $S_1 \subset S'$, $S_1 \triangleleft H$ such that $\mathcal{S}(S_1)$ is coherent, and $S_2 \triangleleft S_1$ such that S_1/S_2 is a chief factor of H, and we assume that $\mathcal{S}(S_2)$ is not coherent. Therefore there is a character $\psi \in \mathcal{S}(S_2)$ such that

$$d|S/S_1|(|Q_1| - 1) = \sum_{\chi \in \mathcal{S}(S_1)} \chi(1)^2 \leq 2d\psi(1).$$

Let $Z/S_2 = Z(S/S_2)$. Then $S_1 \subset Z$ since S is nilpotent. Also, $\psi(1) = d\varphi(1)$, where $\varphi \in \mathrm{Irr}(Q/S_2)$ and

$$\varphi(1)^2 \leq |S/Z| \cdot |Q_1/Z(Q_1)| \leq |S/S_1| \cdot |Q_1/Z(Q_1)|,$$

which gives

$$|S/S_1|(|Q_1| - 1)^2 \leq 4d^2|Q_1/Z(Q_1)|$$

and

$$|S/S_1| \cdot |Z(Q_1)|(|Q_1| - 2) < 4d^2.$$

But, since d is odd and D acts on Q_1 without fixed points, $|Z(Q_1)| \geq 2d+1$, and so $|S/S_1| < 4d^2/(4d^2 - 1) < 2$, which is a contradiction. $\qquad\square$

From (1) and (2), **we will assume henceforth** that Q_1 is a non-abelian p-group for some odd prime number p.

(3) *Let $1 \neq Z \triangleleft H$ be such that $Z \subset Z(Q_1)$. Then $\mathcal{X} = \mathcal{S} - \mathcal{S}(Z)$ is coherent.*

Proof. We first show that $\mathcal{X}_1 = \mathcal{X} \cap \mathcal{S}(S')$ is coherent. Now $|\mathcal{X}_1| \geq 2$ since $Z \neq 1$, and, by Lemma 2(c), if $\chi \in \mathcal{X}_1$, then $\overline{\chi} \neq \chi$. Let $\mathcal{X}_1 = \{\chi_1, \ldots, \chi_r\}$ with $\chi_1(1) \leq \cdots \leq \chi_r(1)$. Since χ_i is induced to H from a character φ_i of $\mathrm{Irr}(S/S' \times Q_1)$, $\chi_i(1)$ is of the form dp^{k_i} for some integer k_i. For $i > 1$,

$$\sum_{1 \leq j < i} \chi_j(1)^2 = |H/S'| - |H/S'Z| - \sum_{i \leq j \leq r} \chi_j(1)^2.$$

The left-hand side of this equation is divisible by d^2, and the right-hand side by $\varphi_i(1)^2$ since $\varphi_i(1)^2 \leq |Q_1/Z(Q_1)| \leq |Q_1/Z|$ by [Is], Corollary 2.30. Thus,

$$(3.1) \qquad\qquad \chi_i(1)^2 \text{ divides } \sum_{1 \leq j < i} \chi_j(1)^2 \text{ for } 1 < i \leq r.$$

Let k be the largest index such that $\chi_k(1) = \chi_1(1)$. Then $k \geq 2$ since $\overline{\chi_1} \neq \chi_1$ and $\{\chi_1,\ldots,\chi_k\}$ is coherent (Lemma 1(b)). Suppose that $k < r$. By (3.1), for $i > k$,

$$2\chi_1(1)\chi_i(1) < p\chi_1(1)\chi_i(1) \leq \chi_i(1)^2 \leq \sum_{1 \leq j < i} \chi_j(1)^2.$$

It follows from Lemma 1 that $\mathcal{X}_1$ is coherent.

Therefore, to prove (3), we may assume that $S' \neq 1$. As in (2), suppose that $S_1 \subset S'$ is such that $S_1 \lhd H$ and $\mathcal{X} \cap \mathcal{S}(S_1)$ is coherent, and suppose that $S_2 \lhd S_1$ is such that S_1/S_2 is a chief factor of H. Suppose that $\mathcal{X} \cap \mathcal{S}(S_2)$ is not coherent. An element ψ of $\mathcal{X} \cap \mathcal{S}(S_2)$ is of the form $\mathrm{Ind}_Q^H(\lambda\theta)$, with $\lambda \in \mathrm{Irr}(S/S_2)$, $\theta \in \mathrm{Irr}(Q_1)$ and $Z \not\subset \mathrm{Ker}(\theta)$. As $\mathrm{Ind}_Q^H(1_{S/S_2} \cdot \theta) \in \mathcal{X}_1$, it follows that $\chi(1)$ divides $\psi(1)$. By Lemma 1(a), there is a character $\psi \in \mathcal{X} \cap \mathcal{S}(S_2)$ such that

$$d|S/S_1| \cdot |Q_1/Z|(|Z| - 1) = \sum_{\chi \in \mathcal{X} \cap \mathcal{S}(S_1)} \chi(1)^2 \leq 2\chi_1(1)\psi(1).$$

But $\chi_1(1)^2 \leq d^2|Q_1/Z(Q_1)| \leq d^2|Q_1/Z|$ and $\psi(1)^2 \leq d^2|S/S_1| \cdot |Q_1/Z|$ since $S_1/S_2 \subset Z(S/S_2)$. It follows that

$$d^2|S/S_1|^2|Q_1/Z|^2(|Z| - 1)^2 \leq 4d^4|S/S_1| \cdot |Q_1/Z|^2,$$

or

$$|S/S_1|(|Z| - 1)^2 \leq 4d^2.$$

But, since D acts on Z without fixed points, $|Z| \geq 2d + 1$, and so $|S/S_1| \leq 1$, which is a contradiction. $\qquad\square$

(4) Notation. Let $\mathcal{Y} = \mathcal{S}(Q') = \{\eta_1,\ldots,\eta_m\}$. Let $Z = [Q_1, Q_1] \cap Z(Q_1)$; since Q_1 is assumed to be non-abelian, $Z \neq 1$. Let

$$\mathcal{X} = \mathcal{S} - \mathcal{S}(Z) = \{\chi_1,\ldots,\chi_n\} \quad \text{with} \quad \chi_1(1) \leq \cdots \leq \chi_n(1),$$

and let $\chi_i(1) = a_i\chi_1(1)$. Then a_i is an integer (see (3)). Now $\mathcal{X} \cap \mathcal{Y} = \emptyset$; also, we know that $\mathcal{X}$ and $\mathcal{Y}$ are coherent, that is to say, there are elements $e_i \in \pm\mathrm{Irr}(G)$ $(1 \leq i \leq n)$ and $e'_j \in \pm\mathrm{Irr}(G)$ $(1 \leq j \leq m)$ such that

$$\mathrm{Ind}_H^G(\chi_i - a_i\chi_1) = e_i - a_ie_1 \quad (2 \leq i \leq n)$$

and

$$\mathrm{Ind}_H^G(\eta_j - \eta_1) \quad = e'_j - e'_1 \quad (2 \leq j \leq m).$$

(5) *For all indices i and j, $(e_i, e'_j) = 0$.*

Proof. Since Ind_H^G is an isometry on $\mathbf{Z}[\mathcal{S}]^\circ$, $(e_i - a_ie_1, e'_2 - e'_1) = 0$ for $2 \leq i \leq n$. Let $\lambda = (e_i - a_ie_1, e'_1) = (e_i - a_ie_1, e'_2)$. If $\lambda \neq 0$, then $a_i = 1$, $\lambda = \pm 1$ and

$e_i - e_1 = \lambda(e'_1 + e'_2)$. But $e_i(1) - e_1(1) = 0$ and $e'_1(1) - e'_2(1) = 0$, whence $e'_1(1) = 0$, which is a contradiction. Consequently, $\lambda = 0$, $e_i \neq \pm e'_1$ for $2 \leq i \leq n$ and $e_1 \neq \pm e'_1$. $\qquad\square$

(6) *Let* $\chi_1(1) = ad$. *Then there is an element* $v \in \mathbf{Z}[\mathrm{Irr}(G)]$ *and an integer* $\lambda \in \mathbf{Z}$ *such that*

$$\mathrm{Ind}_H^G(\chi_1 - a\eta_1) = -ae'_1 + \lambda \sum_{i=1}^{m} e'_i + v,$$

and

$$(v, e'_i) = 0 \quad (1 \leq i \leq m).$$

If λ *is divisible by* a, *then* $\mathcal{S}$ *is coherent.*

Proof. The first assertion follows from the fact that

$$(\mathrm{Ind}_H^G(\chi_1 - a\eta_1), \mathrm{Ind}_H^G(\eta_i - \eta_1)) = a$$

for $i > 1$. Suppose that $\lambda = ax$, $x \in \mathbf{Z}$. Then

$$1 + a^2 = (\chi_1 - a\eta_1, \chi_1 - a\eta_1) = (v, v) + a^2(x - 1)^2 + (m - 1)x^2 a^2,$$

and so $(x - 1)^2 + (m - 1)x^2 \leq 1 + 1/a^2$, and $a > 1$ since $\mathcal{X} \cap \mathcal{Y} = \emptyset$. Thus, $x = 0$, or $x = 1$ and $m = 2$. The second case reduces to the first on replacing e'_1 and e'_2 by $-e'_2$ and $-e'_1$ respectively. We may thus assume that $x = 0$, whence $\lambda = 0$, $(v, v) = 1$ and $\mathrm{Ind}_H^G(\chi_1 - a\eta_1) = v - ae'_1$. Then

$$-1 = (\mathrm{Ind}_H^G(\chi_2 - \chi_1), \mathrm{Ind}_H^G(\chi_1 - a\eta_1)) = (e_2, v) - (e_1, v),$$

and so $v = e_1$ or $v = -e_2$. In the second case, $n = 2$ for, if not,

$$-a_3 = (\mathrm{Ind}_H^G(\chi_3 - a_3\chi_1), \mathrm{Ind}_H^G(\chi_1 - a\eta_1)) = 0,$$

and we can reduce to the first case on replacing e_1 and e_2 by $-e_2$ and $-e_1$ respectively. Thus, $\mathrm{Ind}_H^G(\chi_1 - a\eta_1) = e_1 - ae'_1$. Therefore $\mathcal{X} \cup \mathcal{Y}$ is coherent since Ind_H^G coincides with the isometry $\chi_i \longmapsto e_i$, $\eta_j \longmapsto e'_j$ on $\mathbf{Z}[\mathcal{X}]^\circ$, on $\mathbf{Z}[\mathcal{Y}]^\circ$ and on $\chi_1 - a\eta_1$, and they generate $\mathbf{Z}[\mathcal{X} \cup \mathcal{Y}]^\circ$.

Let $\mathcal{X}_1 = \mathcal{X} \cap \mathcal{S}(S')$ and let $\psi \in \mathcal{S}(S') - (\mathcal{X}_1 \cup \mathcal{Y})$. As in (3.1), $\psi(1)^2$ divides $\sum_{\chi \in \mathcal{X}_1} \chi(1)^2 = |H/S'| - |H/S'Z|$, whence

$$2\eta_1(1)\psi(1) < p\eta_1(1)\psi(1) \leq \psi(1)^2 \leq \sum_{\chi \in \mathcal{X}_1} \chi(1)^2 < \sum_{\chi \in \mathcal{X}_1 \cup \mathcal{Y}} \chi(1)^2.$$

It follows from Lemma 1(a) that $\mathcal{S}(S')$ is coherent, and, from (2), that $\mathcal{S}$ is coherent. $\qquad\square$

(7) *Let $\psi \in \mathrm{Irr}(G)$ be such that ψ is constant on $Z^{\#}$. If $z \in Z^{\#}$, then*

$$\psi(z) \equiv \psi(1) \pmod{|Q|}.$$

Proof. Note that $\psi(z) \in \mathbf{Z}$ since ψ is constant on $Z^{\#}$. Let $\mathcal{K}_s$ be the conjugacy classes of G ($s = 0, 1, 2, \ldots$), and let K_s be the sum of the elements of $\mathcal{K}_s$ in the group algebra of G. Let ω be the homomorphism from $Z(\mathbf{C}[G])$ to $\mathbf{C}$ associated with ψ, i.e., $\omega(K_s) = \psi(K_s)/\psi(1)$. Then

$$(7.1) \qquad \omega(K_i)\omega(K_j) = \sum_s a_{ijs}\omega(K_s), \quad \text{where} \quad K_i K_j = \sum_s a_{ijs} K_s.$$

Let i and j be indices such that $\mathcal{K}_i \cap Z^{\#} \neq \emptyset$ and $\mathcal{K}_j \cap Z^{\#} \neq \emptyset$. Using the notation

$$x \equiv y \pmod{|Q|}$$

if x, y and $\dfrac{x-y}{|Q|}$ are algebraic integers, we show that

$$(7.2) \qquad \psi(1)\omega(K_i)\omega(K_j) \equiv \sum_{s,\, \mathcal{K}_s \cap Z \neq \emptyset} \psi(1)a_{ijs}\omega(K_s) \pmod{|Q|}.$$

Proof. Suppose that $\mathcal{K}_s \cap Z = \emptyset$. Let $u \in \mathcal{K}_i$ and $v \in \mathcal{K}_j$ be such that $uv \in \mathcal{K}_s$, and let $w \in Q^{\#}$. If $u^w = u$ and $v^w = v$, then $u \in C_G(w) \subset H$ and $v \in C_G(w) \subset H$, whence $u \in \mathcal{K}_i \cap H$ and $v \in \mathcal{K}_j \cap H$. Since $Z \triangleleft H$ and Q has trivial intersections in G, $\mathcal{K}_i \cap H \subset Z$ and $\mathcal{K}_j \cap H \subset Z$, whence $uv \in Z$, contrary to hypothesis. It follows that Q acts without fixed points on the set $\{(u,v) \mid u \in \mathcal{K}_i, v \in \mathcal{K}_j, uv \in \mathcal{K}_s\}$ which has cardinality $a_{ijs}|\mathcal{K}_s|$. So, if $x \in \mathcal{K}_s$, then $|Q|$ divides $a_{ijs}|\mathcal{K}_s|\psi(x) = \psi(1)a_{ijs}\omega(K_s)$. This concludes the proof. $\qquad\square$

Let $\mathcal{K}_0 = \{1\}$. When $\mathcal{K}_s \cap Z^{\#} \neq \emptyset$, $\alpha = \omega(K_s)$ is independent of s by hypothesis. The congruence (7.2) then gives

$$(7.3) \qquad \psi(1)\alpha^2 \equiv \psi(1)(a_{ij0} + a_{ij}\alpha) \pmod{|Q|}, \quad \text{with} \quad a_{ij} \in \mathbf{N}.$$

Since d is odd, we may assume that $\mathcal{K}_1 \cap Z^{\#} \neq \emptyset$ and $\mathcal{K}_2 = (\mathcal{K}_1)^{-1}$. Applying (7.3) to $(i,j) = (1,1)$ and then to $(i,j) = (1,2)$, we obtain

$$(7.4) \qquad \psi(1)a_{11}\alpha \equiv \psi(1)(|G:Q| + a_{12}\alpha) \pmod{|Q|}.$$

Applying this to $\psi = 1_G$, we see that

$$|G:Q|a_{11} \equiv |G:Q| + a_{12}|G:Q| \pmod{|Q|}$$

whence $a_{11} \equiv 1 + a_{12} \pmod{|Q|}$ since Q is a Hall subgroup of G; thus, (7.4) becomes

$$\psi(1)(1 + a_{12})\alpha \equiv \psi(1)(|G:Q| + a_{12}\alpha) \pmod{|Q|}$$

whence $\psi(1)\alpha \equiv \psi(1)|G:Q| \pmod{|Q|}$. However, $\alpha = |G:Q|\psi(z)/\psi(1)$ if $z \in Z^{\#}$; therefore, $|G:Q|\psi(z) \equiv |G:Q|\psi(1) \pmod{|Q|}$ and it follows that $\psi(z) \equiv \psi(1) \pmod{|Q|}$. $\qquad\square$

(8) *Conclusion.*

From (5) and (6) we obtain

$$(\mathrm{Res}_H^G e_1', \chi_i - a_i\chi_1) = 0 \quad \text{for} \quad i \geq 2$$

and

$$(\mathrm{Res}_H^G e_1', \chi_1 - a\eta_1) = \lambda - a,$$

whence

$$\mathrm{Res}_H^G e_1' = (\lambda + a\mu) \sum_{i=1}^n a_i\chi_i + \chi',$$

where χ' is a character of H such that $(\chi', \chi_i) = 0$ for all indices i and where $\mu = (\mathrm{Res}_H^G e_1', \eta_1) - 1$. But $\sum_{i=1}^n a_i\chi_i = \dfrac{1}{da}(\rho_H - \rho_{H/Z})$, where ρ_X denotes the character of the regular representation of the group X. On the other hand, $Z \subset \mathrm{Ker}\,\chi'$ since no irreducible component of χ' belongs to $\mathcal{X}$. It follows that, for $z \in Z^\#$,

$$e_1'(z) - e_1'(1) = (\lambda + a\mu)\left(-\frac{|H|}{da}\right) = -|Q|\left(\frac{\lambda}{a} + \mu\right).$$

By (7), a then divides λ and the conclusion follows from (6).   □

Remark. If d is even and $|\mathcal{S}| \geq 2$, the conclusion of the theorem is still valid. Indeed, in this case, Q_1 is abelian, and $|Q_1| > d + 1$ or $S \neq 1$. We can then see that $\mathcal{S}(Q') = \mathcal{S}(S')$ is coherent and, if $\mathcal{S}$ is not coherent, then $S' \neq S$ and $|S/S'| \cdot |Q_1|(|Q_1| - 2) < 4d^2$ (see (2)). But $|Q_1| \geq d + 1$, whence $|S/S'| < 4d^2/(d^2 - 1) < 6$, which is impossible as the case $|S/S'| = 4$ is excluded by the hypothesis that $(|D|, |Q|) = 1$.

References

[B] Bender, H. (1971) 'Transitive Gruppen gerader Ordnung, in denen jede Involution genau einen Punkt festlässt', *J. Algebra* **17**, 527–554.

[F] Feit, W. (1960) 'On a class of doubly transitive permutation groups', *Illinois J. Math.* **4**, 170–186.

[Ha] Hall, M. (1959) *The theory of groups*, Macmillan, New York.

[HKS] Hering, C.; Kantor, W. M.; Seitz, G. M. (1972) 'Finite groups with a split BN-pair of rank 1. I', *J. Algebra* **20**, 435–475.

[Hi] Higman, G. (1963) 'Suzuki 2-groups', *Illinois J. Math.* **7**, 79–96.

[H] Huppert, B. (1967) *Endliche Gruppen*. I, Grundlehren Math. Wiss. **134**, Springer, Berlin.

[HB] Huppert, B.; Blackburn, N. (1982) *Finite groups*. II, Grundlehren Math. Wiss. **242**, Springer, Berlin.

[Is] Isaacs, I. M. (1976) *Character theory of finite groups*, Academic Press, New York.

[KS] Kantor, W. M.; Seitz, G. M. (1972) 'Finite groups with a split BN-pair of rank 1. II', *J. Algebra* **20**, 476–494.

[P] Peterfalvi, T. (1986) 'Le théorème de Bender-Suzuki. I, II', *Astérisque* **142–143**, 141–233, 235–295.

[Si] Sibley, D. A. (1976) 'Coherence in finite groups containing a Frobenius section', *Illinois J. Math.* **20**, 434–442.

[S1] Suzuki, M. (1964) 'On a class of doubly transitive groups. II', *Ann. of Math.* (2) **79**, 514–589.

[S2] Suzuki, M. (1986) *Group theory*. II, Grundlehren Math. Wiss. **248**, Springer, Berlin.

[Z] Zassenhaus, H. (1936) 'Über endliche Fastkörper', *Abhandlungen Math. Sem. Hamburg Univ.* **11**, 187–221.

Index

For EU product safety concerns, contact us at Calle de José Abascal, 56–1°,
28003 Madrid, Spain or eugpsr@cambridge.org.